STUDENT'S SOLUTIONS MANUAL

VOLUME 2: CHAPTERS 17–30

THIRD EDITION

college physics

a strategic approach

knight • jones • field

LARRY K. SMITH
Snow College

PEARSON

Boston Columbus Indianapolis New York San Francisco Hoboken
Amsterdam Cape Town Dubai London Madrid Milan Munich Paris Montréal Toronto
Delhi Mexico City São Paulo Sydney Hong Kong Seoul Singapore Taipei Tokyo

Executive Editor: Becky Ruden
Project Manager: Martha Steele
Managing Development Editor: Cathy Murphy
Senior Development Editor: Alice Houston
Editorial Assistant: Sarah Kaubisch
Compositor: PreMediaGlobal, Inc.
Cover Designer: Tandem Creative, Inc. / Seventeenth Street Studios
Manufacturing Buyer: Jeff Sargent
Vice-President of Marketing: Christy Lesko
Marketing Manager: Will Moore
Senior Market Development Manager: Michelle Cadden

Cover Photo Credit: Franklin Kappa

15

ISBN 10: 0-321-90885-6
ISBN 13: 978-0-321-90885-8

PEARSON www.pearsonhighered.com

Table of Contents

Preface

This *Student's Solutions Manual* is intended to provide you with examples of good problem-solving techniques and strategies. To achieve that, the solutions presented here attempt to:

- Follow, in detail, the problem-solving strategies presented in the text.
- Articulate the reasoning that must be done before computation.
- Illustrate how to use drawings effectively.
- Demonstrate how to utilize graphs, ratios, units, and the many other "tactics" that must be successfully mastered and marshaled if a problem-solving strategy is to be effective.
- Show examples of assessing the reasonableness of a solution.
- Comment on the significance of a solution or on its relationship to other problems.

We recommend you try to solve each problem on your own before you read the solution. Simply reading solutions, without first struggling with the issues, has limited educational value.

As you work through each solution, make sure you understand how and why each step is taken. See if you can understand which aspects of the problem made this solution strategy appropriate. You will be successful on exams not by memorizing solutions to particular problems but by coming to recognize which kinds of problem-solving strategies go with which types of problems.

We have made every effort to be accurate and correct in these solutions. However, if you do find errors or ambiguities, we would be very grateful to hear from you.

WAVE OPTICS

Q17.1. Reason: As a light wave passes from air to glass, every wave in air will create a wave in glass. As a result, the frequency of the wave is the same in air and glass.

Assess: As light waves travel from air to another medium, the frequency stays the same $(f_n = f_{air})$, the wavelength decreases $(\lambda_n = \lambda_{air}/n)$, and the speed at which light travels through the new medium decreases $(v_n = c/n)$.

Q17.3. Reason: The wavelength changes $(\lambda_n = \lambda_{air}/n)$ when the light goes from one medium to another, but the frequency stays the same. Since the color doesn't change, then it must be associated with the frequency.

Assess: Energy of the light is directly related to the frequency as well.

Q17.7. Reason: The fringe separation for the light intensity pattern of a double slit is determined by $\Delta y = \lambda L/d$. We can answer this question by inspecting this relationship.

(a) If λ decreases, Δy will decrease.

(b) If d decreases, Δy will increase.

(c) If L decreases, Δy will decrease.

(d) Since the dot is in the $m = 2$ bright fringe, the path length difference from the two slits is $2\lambda = 1000$ nm.

Assess: The ability to inspect a relationship and answer "what if" questions is a skill that physics will help you develop.

Q17.9. Reason: The width of the central maximum in the intensity pattern for a single slit is given by $w = 2L\lambda/a$. If the wavelength λ is equal to or greater than the slit with a, then w will be equal to or greater than $2L$. It is obvious that $w \ll 2L$, hence $\lambda \ll a$.

Assess: The ability to inspect a relationship and answer questions of this type with confidence is a skill that physics will help you develop.

Q17.13. Reason: The reflection from the left surface of the film is inverted while the reflection from the right surface of the film is not. From the diagram we see that the thickness of the film is 2.5λ. The wave that goes into the film and reflects off the right side will travel 5λ farther. This would normally have produced constructive interference with the wave that reflected from the first (left) surface, but since it was inverted on reflection we will instead get destructive interference.

Assess: The path length difference isn't all that matters here; we must also account for any inversions (phase shifts) that occur.

Q17.17. Reason: The index of refraction of the film is still greater than the index for air, so there is still a phase change at the first boundary. However, if $n_{film} > n_{glass}$ then there will no longer be a phase change at the second boundary. That is, there will be only one phase change instead of two. Therefore, if the conditions were right for destructive interference for two phase changes, then we would have constructive interference for one phase change. Hence, the film would no longer serve as an antireflection coating.

Assess: For the given conditions (the given λ and t) the new film will be an anti-transmission coating and will enhance reflection.

Q17.19. Reason: As light waves travel from air to another medium, the frequency stays the same ($f_n = f_{air}$), the wavelength decreases ($\lambda_n = \lambda_{air}/n$) and the speed at which light travels through the new medium decreases ($v_n = c/n$). Only the frequency of the light remains the same. The correct choice is B.

Assess: Since every disturbance in air creates a disturbance in the second medium, the rate of occurrence of the disturbances (i.e., frequency) in air and the second medium is the same.

Q17.23. Reason: The relationship between the diffraction grating spacing (d), the angle at which a particular order of constructive interference occurs (θ_m), and the wavelength of the light and the order of the constructive interference (m) is $d \sin \theta_m = m\lambda$ and $N = 1/d$.

The relationship between the location of a particular interference order (y_m), the angle at which a particular order of constructive interference occurs (θ_m), and the distance between the grating and the interference pattern viewing screen (L) is $y_m = L \tan \theta_m$.

Use the second expression to determine the angle (θ_1).

$$\theta_1 = \tan^{-1}(y_1/L) = \tan^{-1}(1.9 \times 10^{-2}\ \text{m}/0.80\ \text{m}) = 0.02374\ \text{radian}$$

Use the first expression to determine the number of lines per mm.

$$N = 1/d = \sin \theta_1/\lambda = \sin(0.02374)/5.9 \times 10^{-7}\ \text{m} = 4.0 \times 10^4/\text{m} = 40/\text{mm}$$

The correct choice is B.

Assess: If you check with any vendor selling diffraction gratings, you will find that this is a reasonable number.

Q17.25. Reason: To change from constructive interference to destructive interference the path length difference needs to be changed by a half wavelength. But since the save goes both directions through the film after reflection the film actually only needs to be a quarter wavelength thicker to produce the half wavelength path length difference. The correct choice is D.

Assess: We are talking about the wavelength *in* the film.

Problems

P17.1. Prepare: Light is an electromagnetic wave, and its speed in a certain material is given by Equation 17.1.
Solve: (a) The time light takes is

$$t = \frac{3.0\ \text{mm}}{v_{glass}} = \frac{3.0 \times 10^{-3}\ \text{m}}{c/n} = \frac{3.0 \times 10^{-3}\ \text{m}}{(3.0 \times 10^8\ \text{m/s})/1.50} = 1.5 \times 10^{-11}\ \text{s}$$

(b) The thickness of water is

$$d = v_{water}t = \frac{c}{n_{water}}t = \frac{3.0 \times 10^8\ \text{m/s}}{1.33}(1.5 \times 10^{-11}\ \text{s}) = 3.4\ \text{mm}$$

Assess: In a given time, light travels through a larger thickness of water compared to glass because the refractive index of glass is more than that of water. Please see Table 17.1 for typical indices of refraction.

P17.5. Prepare: This is a straightforward problem if we use the speed of light in water rather than in air.

$$v = \frac{c}{n} = \frac{3.0 \times 10^8\ \text{m/s}}{1.33} = 2.25 \times 10^8\ \text{m/s}$$

We are also given $\Delta x = 150\ \text{m}$.

Solve:

$$\Delta t = \frac{\Delta x}{v} = \frac{150 \text{ m}}{2.25 \times 10^8 \text{ m/s}} = 66.5 \times 10^{-8}\text{ s} = 6.65 \times 10^{-7}\text{ s} \approx 670 \text{ ns}$$

Assess: This is about one-third more than it would take light to travel 150 m in air (remember that light travels about a foot per nanosecond in air), which is what we expect.

P17.7. Prepare: Two closely spaced slits produce a double-slit interference pattern given by Equation 17.6. The interference pattern looks like the photograph of Figure 17.9. It is symmetrical with the $m = 2$ fringes on both sides of and equally distant from the central maximum.

Solve: The bright fringes occur at angles θ_m such that

$$d \sin \theta_m = m\lambda \qquad m = 0, 1, 2, 3, \ldots$$

$$\Rightarrow \sin \theta_2 = \frac{2(500 \times 10^{-9}\text{ m})}{(50 \times 10^{-6}\text{ m})} = 0.02 \Rightarrow \theta_2 = 0.020 \text{ rad} = 0.020 \text{ rad} \times \frac{180°}{\pi \text{ rad}} = 1.1°$$

Assess: We did expect the angle to be small because the wavelength of light is much smaller than the separation of the two slits.

P17.11. Prepare: Two closely spaced slits produce a double-slit interference pattern. The interference pattern looks like the photograph of Figure 17.9. Note that there are 10 gaps between a span of 11 fringes or a distance of 52 mm.

Solve: The formula for the fringe spacing is

$$\Delta y = \frac{\lambda L}{d} \Rightarrow \left(\frac{52 \times 10^{-3}\text{ m}}{10} \right) = \frac{(633 \times 10^{-9}\text{ m})(3.0\text{ m})}{d} \Rightarrow d = 0.37 \text{ mm}$$

Assess: This is a reasonable distance between the slits, ensuring $d/L = 1.22 \times 10^{-4} \ll 1$.

P17.13. Prepare: The bright fringes are where constructive interference takes place. For the $m = 1$ bright fringe just next to the central maximum the path length difference from the two slits is exactly one wavelength.

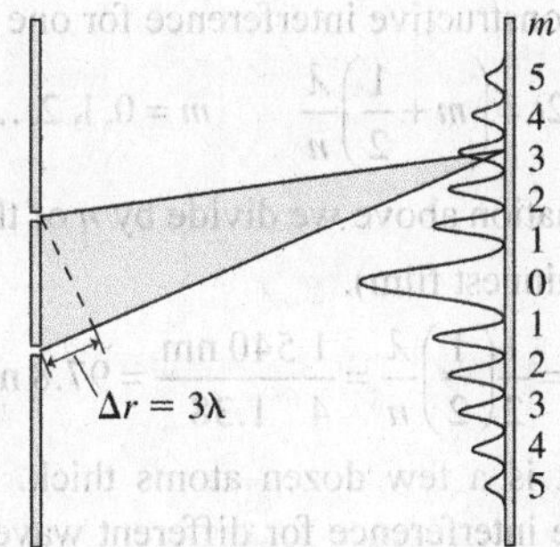

Solve: For $m = 3$ the path length difference is 3λ. The distance from the farther fringe is just 3λ more than the distance from the nearer fringe.

$$2{,}000{,}198.2\lambda + 3\lambda = 2{,}000{,}201.2\lambda$$

Assess: The path length difference is 3λ for $m = 3$ regardless of the distance from one slit to the bright fringe.

P17.15. Prepare: A diffraction grating produces an interference pattern. The interference pattern looks like the diagram in Figure 17.11. The slit spacing for the grating is $d = 1.0 \text{ cm}/1000$.

Solve: The bright constructive-interference fringes are given by Equation 17.12.

$$d \sin \theta_m = m\lambda \qquad m = 0, 1, 2, 3, \ldots$$

$$\Rightarrow \sin \theta_1 = \frac{(1)(550 \times 10^{-9}\text{ m})}{(1.0 \times 10^{-2}\text{ m})/1000} = 0.055 \Rightarrow \theta_1 = 3.15°$$

Likewise, $\sin \theta_2 = 0.110$ and $\theta_2 = 6.32°$. The two angles will be reported to two significant figures as $3.2°$ and $6.3°$.

P17.21. Prepare: The first reflection occurs at the front surface of the MgF_2 film. Here, the index of refraction increases from that of air $(n=1)$ to that of the film $(n=1.38)$, so there will be a reflective phase change. The light then reflects from the rear surface of the film. The index of refraction again increases from 1.38 to 1.50 for the glass, and there is a reflective phase change. With two phase changes, Tactics Box 17.1 tells us that we should use Equation 17.14.

Solve: The film thickness that causes constructive interference at wavelength λ is given by Equation 17.14,

$$\lambda_C = \frac{2nt}{m} \Rightarrow t = \frac{\lambda_C m}{2n} = \frac{(600\times10^{-9}\,\text{m})(1)}{(2)(1.38)} = 217\,\text{nm}$$

where we have used $m=1$ to calculate the thinnest film.

Assess: The film thickness is much less than the wavelength of visible light.

P17.23. Prepare: The film's index of refraction $(n_{\text{film}}=1.60)$ is greater than that of glass $(n_{\text{glass}}=1.50)$, so there will be a reflective phase change at the first boundary (air–film), but not at the second boundary (film–glass). Equation 17.15 gives the relationship for constructive interference for one reflective phase change.

$$2t = \left(m+\frac{1}{2}\right)\frac{\lambda}{n} \qquad m = 0, 1, 2, \ldots$$

We are given $\lambda_{\text{air}}=550$ nm, but in the equation above we divide by n of the film to get the wavelength in the film.

Solve: Solve for t where $m=0$ (for the thinnest film).

$$t = \frac{1}{2}\left(\frac{1}{2}\right)\frac{\lambda}{n} = \frac{1}{4}\frac{550\,\text{nm}}{1.60} = 85.9\,\text{nm}$$

Assess: The answer seems reasonable; it is a few dozen atoms thick. We used three significant figures as indicated by the given data.

P17.27. Prepare: Because you happen to remember the index of refraction of gasoline $(n_{\text{gas}}=1.38)$ is greater than that of water $(n_{\text{water}}=1.33)$, you realize there will be a reflective phase change at the first boundary (air–gasoline), but not at the second boundary (gasoline–water). Equation 17.15 gives the relationship for constructive interference for one reflective phase change.

$$2t = \left(m+\frac{1}{2}\right)\frac{\lambda}{n} \qquad m = 0, 1, 2, \ldots$$

We are given $\lambda_{\text{air}}=540$ nm, but in the equation above we divide by n of the film to get the wavelength in the film.

Solve: Solve for t where $m=0$ (for the thinnest film).

$$t = \frac{1}{2}\left(\frac{1}{2}\right)\frac{\lambda}{n} = \frac{1}{4}\frac{540\,\text{nm}}{1.38} = 97.8\,\text{nm}$$

Assess: The answer seems reasonable; it is a few dozen atoms thick. Other thicknesses would emphasize other colors (that is, there would be constructive interference for different wavelengths). We used three significant figures as indicated by the given data.

P17.29. Prepare: Since there is only one slit in the aluminum foil we know this is a case of single-slit diffraction. We could derive the answer from a couple of different formulae, but the textbook already does the work and gives the width of the central maximum for a single-slit diffraction pattern in Equation 17.19, $w=2\lambda L/a$.

We are given $\lambda=680$ nm, $L=5.5$ m, and $w=0.080$ m.

Solve: Solve Equation 17.20 for a, which is the width of the slit we are asked for.

$$a = \frac{2\lambda L}{w} = \frac{2(680\times10^{-9}\,\text{m})(5.5\,\text{m})}{0.080\,\text{m}} = 9.4\times10^{-5}\,\text{m} = 94\,\mu\text{m}$$

Assess: All of these numbers are reasonable. The slit is thin, as we would expect.

P17.33. Prepare: Light passing through a circular aperture leads to a diffraction pattern that has a circular central maximum surrounded by a series of secondary bright fringes. The width of the central maximum for a circular aperture of diameter D under the small-angle approximation, which is almost always valid for the diffraction of light, is given by Equation 17.22.

Solve: The width is

$$w = \frac{2.44\lambda L}{D} = \frac{(2.44)(500\times10^{-9}\text{ m})(2.0\text{ m})}{0.50\times10^{-3}\text{ m}} = 4.9\text{ mm}$$

P17.35. Prepare: Light passing through a circular aperture leads to a diffraction pattern that has a circular central maximum surrounded by a series of secondary bright fringes. The intensity pattern will look like Figure 17.27. We will use Equation 17.22 to find the distance between the screen and the pinhole.

Solve: From Equation 17.22,

$$L = \frac{Dw}{2.44\lambda} = \frac{(0.12\times10^{-3}\text{ m})(1.0\times10^{-2}\text{ m})}{2.44(633\times10^{-9}\text{ m})} = 78\text{ cm}$$

Assess: A distance of approximately 1 meter between the screen and pinhole is reasonable.

P17.37. Prepare: The speed of light in a material is determined by the refractive index as $v = c/n$. To acquire data from memory, a total time of only 2.0 ns is allowed. This time includes 0.5 ns that the memory unit takes to process a request. Thus, the *travel time* for an infrared light pulse from the central processing unit to the memory unit and back is 1.5 ns.

Solve: Let d be the distance between the central processing unit and the memory unit. The refractive index of silicon for infrared light is $n_{\text{si}} = 3.50$. Then,

$$1.5\text{ ns} = \frac{2d}{v_{\text{Si}}} = \frac{2d}{c/n_{\text{Si}}} = \frac{2dn_{\text{Si}}}{c} \Rightarrow d = \frac{(1.5\text{ ns})c}{2n_{\text{Si}}} = \frac{(1.5\times10^{-9}\text{ s})(3.0\times10^{8}\text{ m/s})}{2(3.5)} \Rightarrow d = 6.4\text{ cm}$$

P17.41. Prepare: Two closely spaced slits produce a double-slit interference pattern. The interference pattern looks like the photograph of Figure 17.9. The bright fringes are located at positions given by Equation 17.6, $d\sin\theta_m = m\lambda$.

Solve: For the $m = 3$ bright orange fringe, the interference condition is $d\sin\theta_3 = 3(600\times10^{-9}\text{ m})$. For the $m = 4$ bright fringe the condition is $d\sin\theta_4 = 4\lambda$. Because the position of the fringes is the same,

$$d\sin\theta_3 = d\sin\theta_4 = 4\lambda = 3(600\times10^{-9}\text{ m}) \Rightarrow \lambda = \frac{3}{4}(600\times10^{-9}\text{ m}) = 450\text{ nm}$$

Assess: The wavelength is in the visible region.

P17.43. Prepare: We plug our specific numbers into Figure 17.11.
We are given $\lambda = 670$ nm. The slit spacing is the inverse of the line count, $d = 1/750\text{ lines/mm} = 1.33\times10^{-6}\text{ m}$.
Using Equation 17.12,

$$\theta_1 = \sin^{-1}\left(\frac{m\lambda}{d}\right) = \sin^{-1}\left(\frac{(1)(670\text{ nm})}{1.33\times10^{-6}\text{ m}}\right) = 0.53\text{ rad}$$

and then Equation 17.13 gives y_1,

$$y_1 = L\tan\theta_1 = (1.0\text{ m})(\tan(0.53\text{ rad})) = 0.58\text{ m}$$

However, when we redo the calculations for $m = 2$ we find that $m\lambda/d > 1$ and so we can't take the arcsine. That is, only one order (besides the central maximum) will appear on the screen.

Solve:

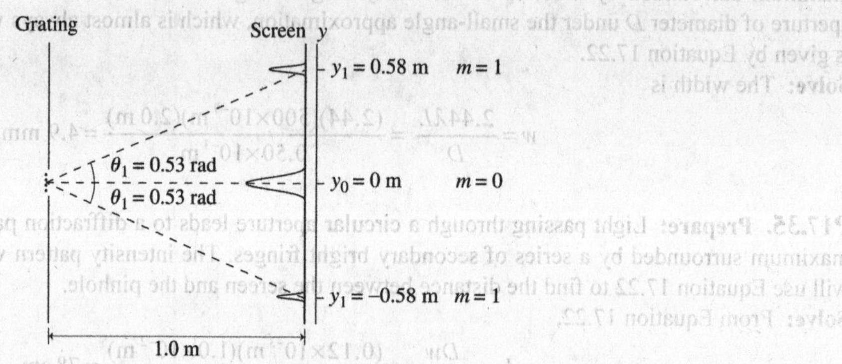

Assess: The values and spacings shown in the sketch are typical; you could verify their reasonableness in a lab.

P17.47. Prepare: A diffraction grating produces an interference pattern that is determined by both the slit spacing and the wavelength used. The visible spectrum spans the wavelengths 400 nm to 700 nm. According to Equation 17.13, the distance y_m from the center to the mth maximum is $y_m = L \tan \theta_m$. The angle of diffraction is determined by the constructive-interference condition $d \sin \theta_m = m\lambda$, where $m = 0, 1, 2, 3, \ldots$ The width of the rainbow for a given fringe order is thus $w = y_{\text{red}} - y_{\text{violet}}$.

Solve: The slit spacing is

$$d = \frac{1\,\text{mm}}{600} = \frac{1.0 \times 10^{-3}\,\text{m}}{600} = 1.6667 \times 10^{-6}\,\text{m}$$

For the red wavelength and for the $m = 1$ order,

$$d \sin \theta_1 = (1)\lambda \Rightarrow \theta_1 = \sin^{-1}\frac{\lambda}{d} = \sin^{-1}\frac{700 \times 10^{-9}\,\text{m}}{1.6667 \times 10^{-6}\,\text{m}} = 24.83°$$

From the equation for the distance of the fringe,

$$y_{\text{red}} = L \tan \theta_1 = (2.0\,\text{m}) \tan (24.83°) = 92.56\,\text{cm}$$

Likewise for the violet wavelength,

$$\theta_1 = \sin^{-1}\left(\frac{400 \times 10^{-9}\,\text{m}}{1.6667 \times 10^{-6}\,\text{m}}\right) = 13.88° \Rightarrow y_{\text{violet}} = (2.0\,\text{m}) \tan (13.88°) = 49.42\,\text{cm}$$

The width of the rainbow is thus $92.56\,\text{cm} - 49.42\,\text{cm} = 43.1\,\text{cm}$ or $43\,\text{cm}$ or 43 cm to two significant figures.

Assess: A width of roughly half a meter seems reasonable.

P17.49. Prepare: A diffraction grating produces an interference pattern that is determined by both the slit spacing and the wavelength used. According to Equation 17.13, the distance y_m from the center to the mth maximum is $y_m = L \tan \theta_m$. The angle of diffraction is determined by the constructive-interference condition $d \sin \theta_m = m\lambda$, where $m = 0, 1, 2, 3, \ldots$ An 800 line/mm diffraction grating has a slit spacing $d = (1.0 \times 10^{-3}\,\text{m})/800 = 1.25 \times 10^{-6}\,\text{m}$.

Solve: Referring to Figure P17.49, the angle of diffraction is given by

$$\tan \theta_1 = \frac{y_1}{L} = \frac{0.436\,\text{m}}{1.0\,\text{m}} = 0.436 \Rightarrow \theta_1 = 23.557° \Rightarrow \sin \theta_1 = 0.400$$

Using the constructive-interference condition $d \sin \theta_m = m\lambda$,

$$\lambda = \frac{d \sin \theta_1}{1} = (1.25 \times 10^{-6}\,\text{m})(0.400) = 500\,\text{nm}$$

We can obtain the same value of λ by using the second-order interference fringe. We first obtain θ_2.

$$\tan\theta_2 = \frac{y_2}{L} = \frac{0.436\text{ m} + 0.897\text{ m}}{1.0\text{ m}} = 1.333 \Rightarrow \theta_2 = 53.12° \Rightarrow \sin\theta_2 = 0.800$$

Using the constructive-interference condition,

$$\lambda = \frac{d\sin\theta_2}{2} = \frac{(1.25\times10^{-6}\text{ m})(0.800)}{2} = 500\text{ nm}$$

Assess: Calculations with the first-order and second-order fringes of the interference pattern give the same value for the wavelength.

P17.51. Prepare: We will assume that the listeners did not hear any other loud spots between the center and 1.4 m on each side, $m = 1$. We'll use the diffraction grating equations. First solve Equation 17.13 for θ_1 and insert it into Equation 17.12.
We need the wavelength of the sound waves, and we'll use the fundamental relationship for periodic waves to get it.

$$\lambda = \frac{v}{f} = \frac{340\text{ m/s}}{10\,000\text{ Hz}} = 3.4\text{ cm} = 0.034\text{ m}$$

We are also given $L = 10$ m and $y_1 = 1.4$ m.

Solve:

$$\theta_1 = \tan^{-1}\left(\frac{y_1}{L}\right) = \tan^{-1}\left(\frac{1.4\text{ m}}{10\text{ m}}\right) = 0.14\text{ rad}$$

(This may be small enough to use the small-angle approximation, but we are almost finished with the problem without it, so maybe we can save the approximation and try it in the assess step.)

$$d = \frac{(1)\lambda}{\sin\theta_1} = \frac{0.034\text{ m}}{\sin(0.14\text{ rad})} = 0.25\text{ m} = 25\text{ cm}$$

Assess: These numbers seem reasonable given the size (wavelength) of sound waves.
With the small-angle approximation, $\theta \approx \sin\theta \approx \tan\theta$,

$$d = \frac{m\lambda}{y_m/L} = \frac{1(0.034\text{ m})}{1.4\text{ m}/10\text{ m}} = 24\text{ cm}$$

This is almost the same, but rounded down to two significant figures rather than up. The θ_1 we computed seemed almost small enough to use the approximation, and it would probably be OK for many applications when the angle is this small.

P17.55. Prepare: A diffraction grating produces an interference pattern. The interference pattern looks like the diagram in Figure 17.11. The angle of diffraction is determined by the constructive-interference condition, Equation 17.12, $d\sin\theta_m = m\lambda$, where $m = 0, 1, 2, 3, \ldots$
Solve: 500 lines per mm on the diffraction grating gives a spacing between the two lines of $d = 1\text{ mm}/500 = (1\times10^{-3}\text{ m})/500 = 2.0\times10^{-6}$ m. The wavelength diffracted at angle $\theta_m = 30°$ in order m is

$$\lambda = \frac{d\sin\theta_m}{m} = \frac{(2.0\times10^{-6}\text{ m})\sin 30°}{m} = \frac{1000\text{ nm}}{m}$$

We're told it is *visible* light that is diffracted at 30°, and the wavelength range for visible light is 400 nm–700 nm. Only $m = 2$ gives a visible light wavelength, so $\lambda = 500$ nm.

P17.57. Prepare: The angle of diffraction is determined by the constructive-interference condition, Equation 17.12, $d\sin\theta_m = m\lambda$, where $m = 0, 1, 2, 3, \ldots$ In this case, however, we do not know m. In fact, we actually have two (very similar) equations with two unknowns: m, and λ, and it is λ we seek (it has the same value in both equations). We solve the system of two equations with two unknowns. We are given $\theta_m = 45.7°$ and $\theta_{m+1} = 72.6°$.
Solve: The two different (but very similar) equations are

$$\lambda = \frac{d}{m}\sin\theta_m \quad \text{and} \quad \lambda = \frac{d}{m+1}\sin\theta_{m+1}$$

There are various methods for solving a system of two equations with two unknowns, but in this case it will be easiest to eliminate λ, solve for m, then plug back in to get λ; we accomplish this by first dividing the two equations.

$$\frac{\lambda}{\lambda}=\frac{\dfrac{d}{m}\sin\theta_m}{\dfrac{d}{m+1}\sin\theta_{m+1}}=1\Rightarrow \frac{m}{m+1}=\frac{\sin\theta_m}{\sin\theta_{m+1}}$$

At this point one could plug in the angles and easily see that $m/(m+1)=0.75=3/4$ and conclude that $m=3$. A more generalized approach would be to continue solving for m with some more steps of algebra,

$$m=\frac{\left(\dfrac{\sin\theta_m}{\sin\theta_{m+1}}\right)}{1-\left(\dfrac{\sin\theta_m}{\sin\theta_{m+1}}\right)}=\frac{\left(\dfrac{\sin 45.7^\circ}{\sin 72.6^\circ}\right)}{1-\left(\dfrac{\sin 45.7^\circ}{\sin 72.6^\circ}\right)}=3$$

We now know that $m=3$, so

$$\lambda=\frac{d}{m}\sin\theta_m=\frac{\frac{1\text{ mm}}{500\text{ slits}}}{3}\sin 45.7^\circ=477\text{ nm}$$

Assess: This wavelength is in the visible range.

P17.63. Prepare: The crack in the cave is like a single slit that causes the ultrasonic sound beam to diffract. We will use Equation 17.17 for complete destructive interference. A schematic representation of the problem is shown.

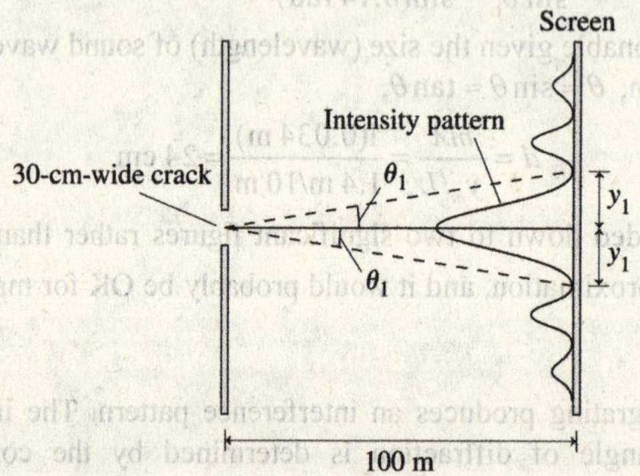

Solve: The wavelength of the ultrasound wave is $\lambda=340\text{ m/s}/30\text{ kHz}=0.0113\text{ m}$.
Using the condition for complete destructive interference with $p=1$,

$$a\sin\theta_1=\lambda \Rightarrow \theta_1=\sin^{-1}\left(\frac{0.0113\text{ m}}{0.30\text{ m}}\right)=2.165^\circ$$

From the geometry of the diagram, the width of the sound beam is
$$w=2y_1=2(100\text{ m}\times\tan\theta_1)=200\text{ m}\times\tan 2.165^\circ=7.6\text{ m}$$

Assess: The small-angle approximation is almost always valid for the diffraction of light, but may not be valid for the diffraction of sound waves, which have a much larger wavelength.

P17.65. Prepare: A narrow slit produces a single-slit diffraction pattern. The diffraction-intensity pattern from a single slit will look like Figure 17.26. As given by Equation 17.19, the dark fringes in this diffraction pattern:
$y_p=p\lambda L/a,\;\;p=1,2,3,\dots$

Solve: We note from Figure P17.65 that the first minimum is 0.50 cm away from the central maximum. Using the above equation, the slit width is

$$a = \frac{p\lambda L}{y_p} = \frac{(1)(500\times10^{-9}\,\text{m})(1.0\,\text{m})}{0.50\times10^{-2}\,\text{m}} = 0.10\,\text{mm}$$

Assess: This is a typical slit width for diffraction.

P17.67. Prepare: Two closely spaced slits produce a double-slit interference pattern with the intensity graph looking like Figure 17.9. The intensity pattern due to a single slit diffraction looks like Figure 17.26. Both the spectra consist of a central maximum flanked by a series of secondary maxima and dark fringes.

Solve: (a) The light intensity shown in Figure P17.69 corresponds to a double-slit aperture. This is because the fringes are equally spaced and the decrease in intensity with increasing fringe order occurs slowly.

(b) From Figure P17.69, the fringe spacing is $\Delta y = 1.0\,\text{cm} = 1.0\times10^{-2}\,\text{m}$. Therefore, using Equation 17.9

$$\Delta y = \frac{\lambda L}{d} \Rightarrow d = \frac{\lambda L}{\Delta y} = \frac{(6.00\times10^{-9}\,\text{m})(2.5\,\text{m})}{0.010\,\text{m}} = 0.15\,\text{mm}$$

Assess: This is a typical value for the spacing between the two slits.

P17.71. Prepare: The laser beam is diffracted through a circular aperture. We will use Equations 17.21 and 17.22.

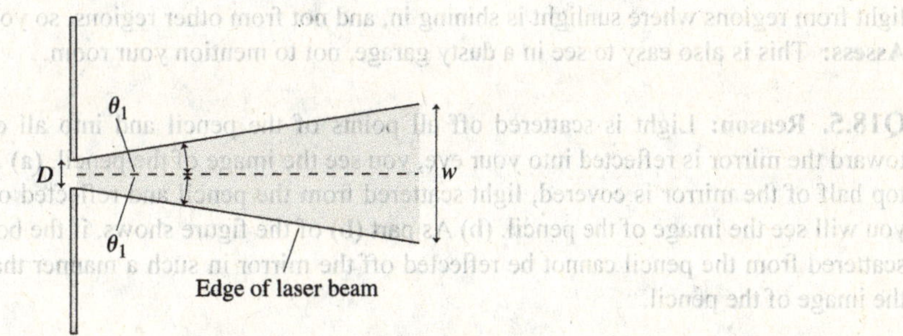

Solve: (a) The laser light emerges through a circular aperture at the end of the laser. This aperture causes diffraction, hence the laser beam must gradually spread out. The diffraction angle is small enough that the laser beam *appears* to be parallel over short distances. But if you observe the laser beam at a large distance it is easy to see that the diameter of the beam is slowly increasing.

(b) The position of the first minimum in the diffraction pattern is more or less the "edge" of the laser beam. For diffraction through a circular aperture, the first minimum is at an angle

$$\theta_1 = \frac{1.22\lambda}{D} = \frac{1.22(633\times10^{-9}\,\text{m})}{0.0010\,\text{m}} = 7.723\times10^{-4}\,\text{rad} = 0.044°$$

(c) The diameter of the laser beam is the width of the diffraction pattern.

$$w = \frac{2.44\lambda L}{D} = \frac{2.44(633\times10^{-9}\,\text{m})(3.0\,\text{m})}{0.0010\,\text{m}} = 4.6\,\text{mm}$$

(d) At $L = 1\,\text{km} = 1000\,\text{m}$, the diameter is

$$w = \frac{2.44\lambda L}{D} = \frac{2.44(633\times10^{-9}\,\text{m})(1000\,\text{m})}{0.0010\,\text{m}} = 1.5\,\text{m}$$

Assess: The above values are quite reasonable.

18

RAY OPTICS

Q18.3. Reason: No, you cannot see the sun's rays on a clear day. Rays of light are only visible if they reach your eye. For that to happen they have to be scattered off of something. When sunlight goes through treetops into a foggy forest there is plenty of water vapor to scatter the light and let it reach your eyes. In that case, your eyes can detect light from regions where sunlight is shining in, and not from other regions, so you can effectively see the rays.
Assess: This is also easy to see in a dusty garage, not to mention your room.

Q18.5. Reason: Light is scattered off all points of the pencil and into all directions of space. If light directed toward the mirror is reflected into your eye, you see the image of the pencil. **(a)** As part **(a)** of the figure shows, if the top half of the mirror is covered, light scattered from the pencil and reflected off the mirror can enter your eye and you will see the image of the pencil. **(b)** As part **(b)** of the figure shows, if the bottom half the mirror is covered, light scattered from the pencil cannot be reflected off the mirror in such a manner that it enters your eye. You cannot see the image of the pencil.

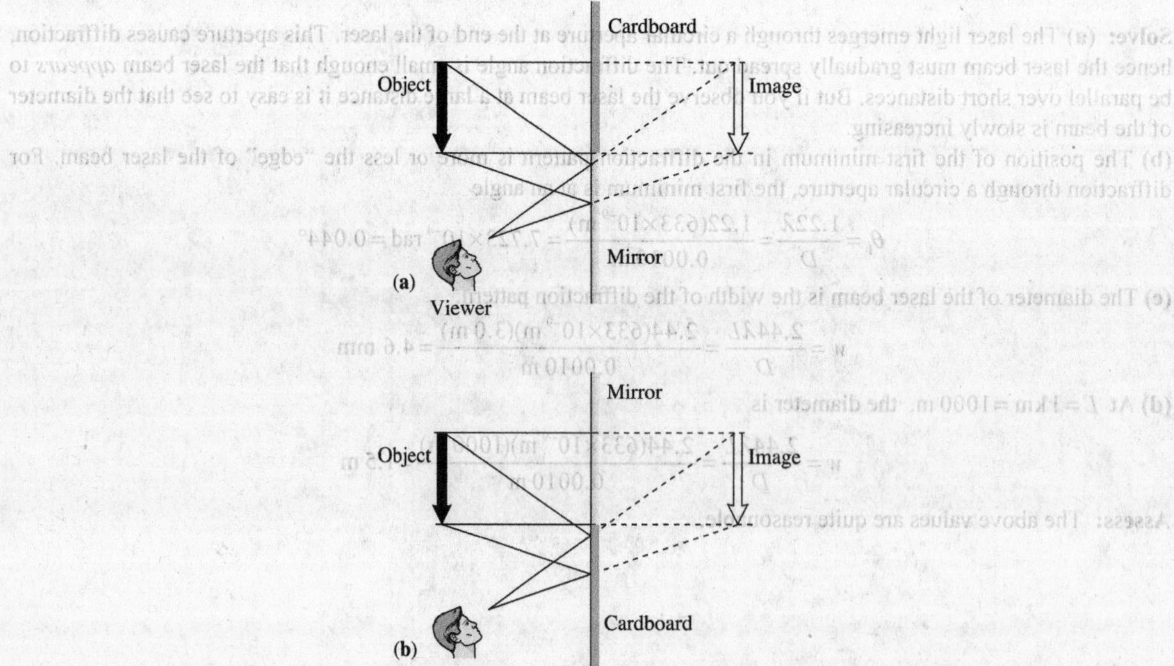

Assess: Since reflection is involved, the key to the question is the knowledge that the angle of reflection is equal to the angle of incidence.

Q18.9. Reason: (a) The image of a point on the object is found by keeping track of light rays leaving that point of the object and refracting through the lens. The image is someplace along the refracted ray. If we consider just one ray, we don't know where the image is—just that it is somewhere along that ray. If we consider a second ray, the image is somewhere along that refracted ray. The only way the image can be along both rays is to be at the point of intersection of the rays. The point of intersection pins down the image of the object. The minimum number of rays needed to locate the image of a point is two.
(b) Light rays are scattered off each point of an object into all directions in space. Only those rays that pass through the lens ultimately form the image. Numerous rays from each point of the object strike the lens and refract to the image point.
Assess: When locating the image of an object, always pick the two simplest rays you can find.

Q18.11. Reason: While the law of refraction depends on the index of the media, the law of reflection does not. Under water the angle of incidence will still be equal to the angle of reflection. There is no reason for a ray to travel a different path in water than in air in this case. Hence, the sun's rays will be focused the same distance from the mirror.
Assess: Of course the preceding analysis is not true for light passing through a lens that might be immersed in either water or air; in that case the index of refraction matters. But with a mirror the water doesn't change anything.

Q18.15. Reason: If we assume the thin plastic wall is of uniform thickness it won't have much effect on the overall effect of the lens. Since the air-filled lens is immersed in water the rays will bend in the opposite directions, relative to the local normal, than in a glass lens in air. Consequently this air-filled lens will actually be a diverging lens, even though in the shape of a regular converging lens.
Assess: You can try this out by inflating a resealable plastic bag and immersing it in water.

Q18.17. Reason: For the reflected ray the angle or reflection is equal to the angle if incidence, hence $\theta_1 = 90° - 50° = 40°$. The correct choice is A.
Assess: For reflection the angle of reflection is equal to the angle of incidence.

Q18.19. Reason: Since the index of refraction of air is less than the index of refraction of glass, total internal reflection will not occur as light travels from air to glass. The correct choice is C.
Assess: If the situation is turned around and the light is traveling from glass to air, total internal reflection can occur.

Q18.21. Reason: You are 2.4 m in front of the mirror and you are photographing an image that appears to be 2.4 m behind the mirror: Adjust the focus of the camera for an object distance of 4.8 m. The correct choice is D.
Assess: The object isn't actually there, it just appears to be there. For example you cannot focus this image onto a screen because it is not real.

Q18.25. Reason: Given the object and image distance, we can determine the focal length by
$$\frac{1}{f} = \frac{1}{s} + \frac{1}{s'} \text{ inserting values obtain } \frac{1}{f} = \frac{1}{12\,\text{cm}} + \frac{1}{36\,\text{cm}} = \frac{1}{9.0\,\text{cm}} \text{ or } f = 9.0\,\text{cm}$$
The correct choice is A.
Assess: Notice that the image is real and the magnitude of the magnification is greater than one. This will occur when the object is between the focal point and the radius of curvature of the lens. As a result, for this case the focal length must be less than the object distance and it is.

Q18.27. Reason: For a convex mirror all images are virtual and erect, and the magnitude of the magnification is less than one. The correct choice is A.

Assess: The side mirrors on cars are convex. The image of the car behind you is erect (it doesn't look like it is traveling down the road upside down), it appears smaller than it actually is (the magnitude of the magnification is less than one), it is virtual (you can't focus it onto a card or screen), and it appears farther away than it actually is since it is so small.

Problems

P18.1. Prepare: This problem employs the fact that rays travel in straight lines.

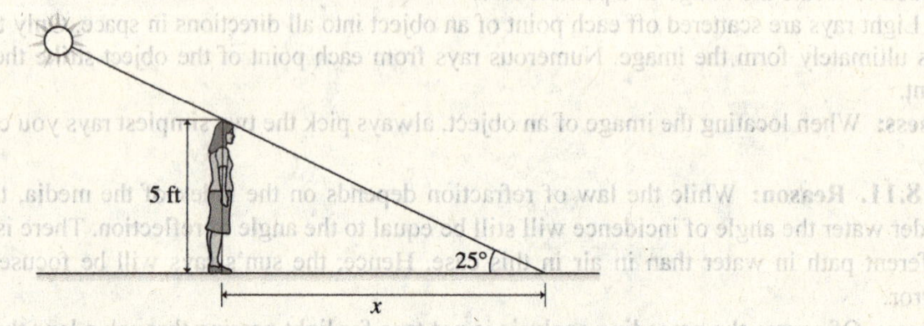

Solve: Remember that for right triangles $\tan\theta = \text{opp/adj}$. In our case $\theta = 25°$ and $\text{opp} = 5$ ft.

$$x = \frac{5\ \text{ft}}{\tan 25°} = 11\ \text{ft}$$

where we have rounded to two significant figures.

Assess: When the sun is less than $45°$ above the horizon we expect the shadow to be longer than the height of the object.

P18.3. Prepare: The light source is a point source and light rays travel in straight lines. We will use geometry to solve this problem.

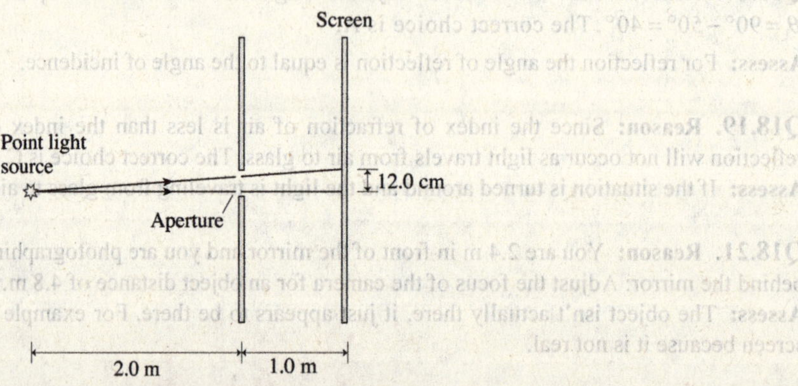

Solve: Let w be the width of the aperture. Then from the geometry of the figure,

$$\frac{w}{2.00\ \text{m}} = \frac{12.0\ \text{cm}}{2.00\ \text{m} + 1.00\ \text{m}} \Rightarrow w = 8.00\ \text{cm}$$

Assess: We expected the aperture's width to be smaller than 12.0 cm, so a value of 8.00 cm is reasonable.

P18.7. Prepare: Use the ray model of light and the law of reflection. We only need one ray of light that leaves your toes and reflects in your eye.

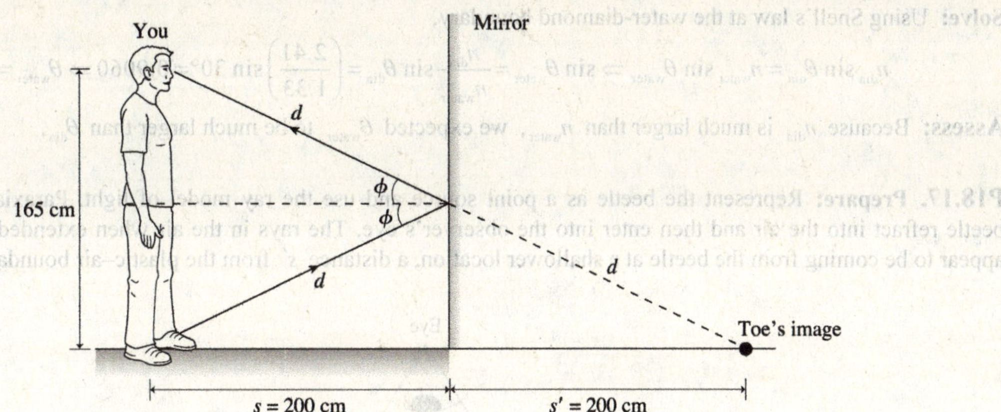

Solve: From the geometry of the diagram, the distance from your eye to the toe's image is

$$2d = \sqrt{(400 \text{ cm})^2 + (165 \text{ cm})^2} = 433 \text{ cm}$$

Assess: The light appears to come from your toe's image.

P18.9. Prepare: Use the ray model of light. The sun is a point source of light. A visual overview of the problem is shown below in which the first four steps of the Tactics Box 18.1 have been identified. A ray that arrives at the diver $50°$ above horizontal is refracted into the water at $\theta_{\text{water}} = 40°$.

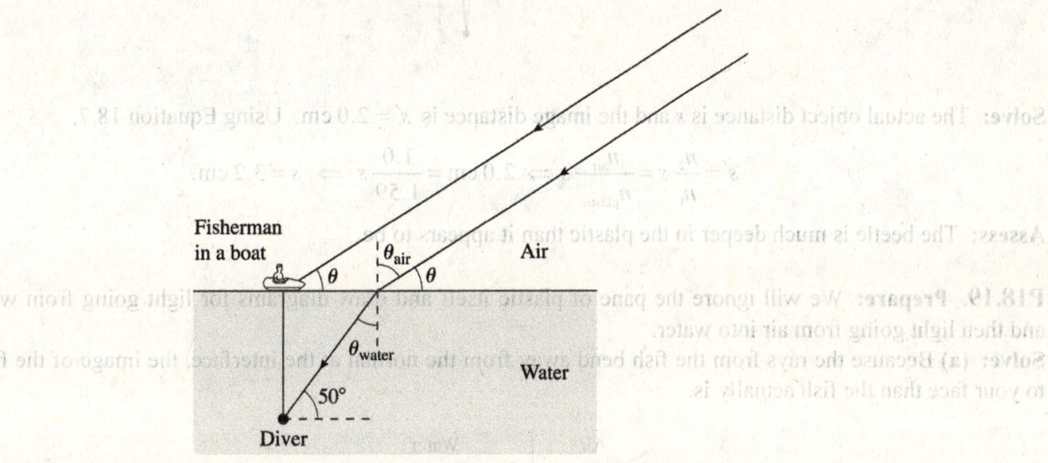

Solve: Using Snell's law at the water-air boundary

$$n_{\text{air}} \sin \theta_{\text{air}} = n_{\text{water}} \sin \theta_{\text{water}} \Rightarrow \sin \theta_{\text{air}} = \frac{n_{\text{water}}}{n_{\text{air}}} \sin \theta_{\text{water}} = \left(\frac{1.33}{1.0}\right) \sin 40° \Rightarrow \theta_{\text{air}} = 58.7°$$

Thus the height above the horizon is $\theta = 90° - \theta_{\text{air}} = 31.3°$. Because the sun is far away from the fisherman (and the diver), the fisherman will see the sun at the same angle of $31°$ above the horizon.

Assess: θ_{air} is larger than θ_{water} as we would have expected.

P18.13. Prepare: Use the ray model of light. The figure incorporates the first four steps of Tactics Box 18.1.

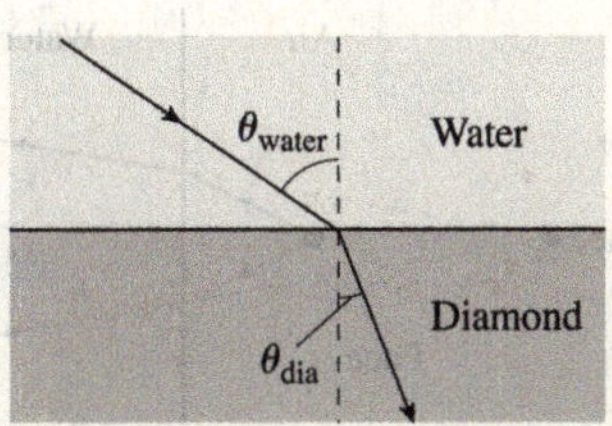

Solve: Using Snell's law at the water-diamond boundary,

$$n_{dia} \sin \theta_{dia} = n_{water} \sin \theta_{water} \Rightarrow \sin \theta_{water} = \frac{n_{dia}}{n_{water}} \sin \theta_{dia} = \left(\frac{2.41}{1.33}\right) \sin 30° = 0.9060 \Rightarrow \theta_{water} = 65°$$

Assess: Because n_{dia} is much larger than n_{water}, we expected θ_{water} to be much larger than θ_{dia}.

P18.17. Prepare: Represent the beetle as a point source and use the ray model of light. Paraxial rays from the beetle refract into the air and then enter into the observer's eye. The rays in the air when extended into the plastic appear to be coming from the beetle at a shallower location, a distance s' from the plastic–air boundary.

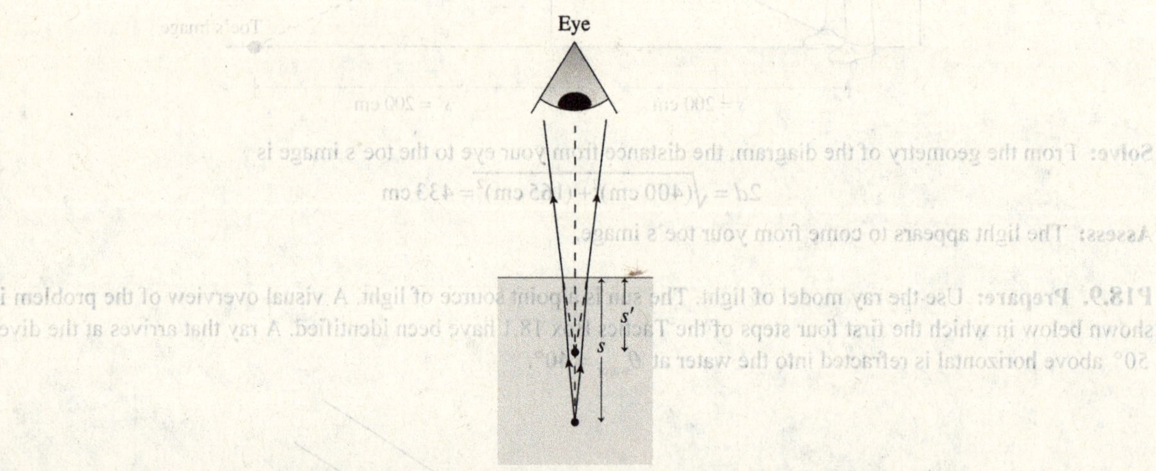

Solve: The actual object distance is s and the image distance is $s' = 2.0$ cm. Using Equation 18.7,

$$s' = \frac{n_2}{n_1} s = \frac{n_{air}}{n_{plastic}} s \Rightarrow 2.0 \text{ cm} = \frac{1.0}{1.59} s \Rightarrow s = 3.2 \text{ cm}.$$

Assess: The beetle is much deeper in the plastic than it appears to be.

P18.19. Prepare: We will ignore the pane of plastic itself and draw diagrams for light going from water into air and then light going from air into water.
Solve: (a) Because the rays from the fish bend away from the normal at the interface, the image of the fish is closer to your face than the fish actually is.

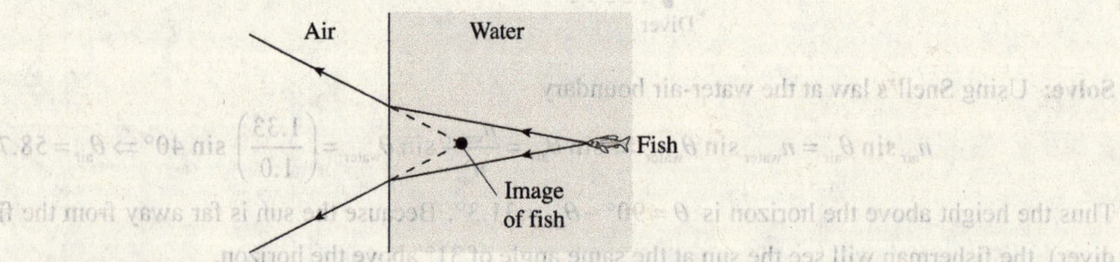

(b) Because the rays from your face bend toward the normal at the interface, the image of your face is farther from the fish than your face actually is.

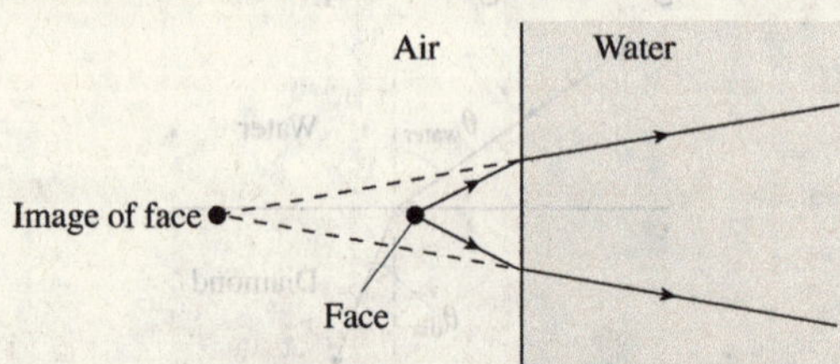

Assess: Only a couple of the many rays were shown in each case. See if you can verify these conclusions next time you go swimming.

P18.23. Prepare: Use ray tracing to locate the image. The figure shows the ray-tracing diagram using the steps of Tactics Box 18.2.
Solve:

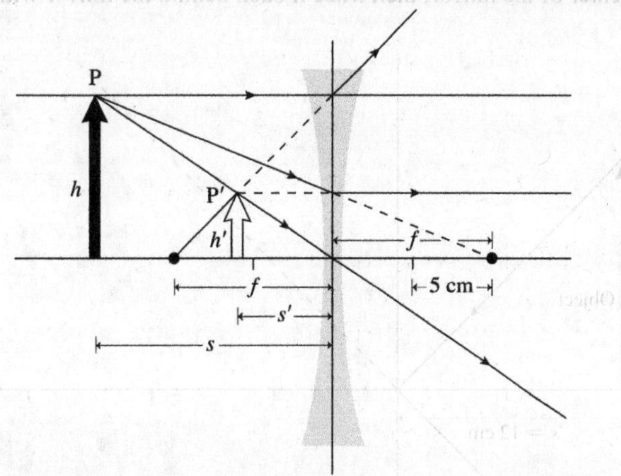

The three rays after refraction do not converge at a point, but they appear to come from P′. P′ is 6 cm from the diverging lens, so $s' = -6.0$ cm. The image is upright and virtual.
Assess: Draw ray tracings accurately to scale.

P18.27. Prepare: Carefully draw this to scale. The object is the tooth. We are given that $m = 2.0$ and $s = 1.2$ cm. We can solve for s' from $m = -s'/s$.

$$s' = -ms = -(2.0)(1.2 \text{ cm}) = -2.4 \text{ cm}$$

This tells us that the image is behind the mirror and virtual.
We also know that the magnification gives the ratio of the image height to the object height, so we can draw scaled arrows representing our object and image.
Given that the image is erect (not inverted) and these numbers we can draw the ray-tracing diagram.
Solve: Draw the object 1.2 cm to the left of the mirror plane and the image twice as tall 2.4 cm to the right of the mirror plane. Draw the three special rays. The light physically stays on the left side of the mirror, so extend the lines as dotted lines behind the mirror to the virtual image.

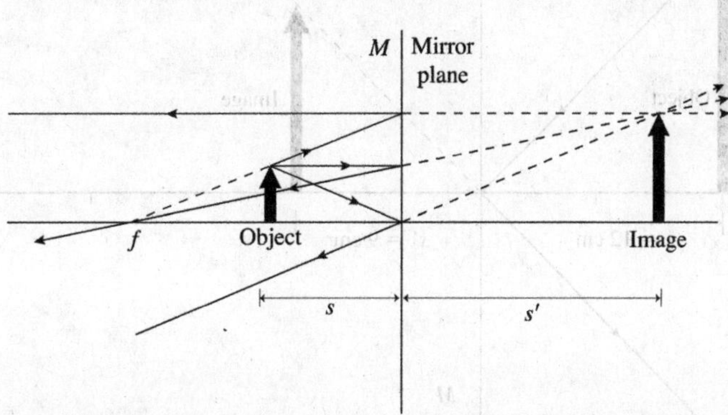

Even though we drew a straight vertical mirror plane, thinking through the two possibilities shows that for the light rays to go where they do the mirror must be concave.
The focal point appears to be twice as far from the mirror as the object, or $f \approx 2.4$ cm.

Assess: The concave make-up mirrors you see in fancy hotels also produce an erect magnified virtual image if the object is within the focal length.

P18.29. Prepare: For a convex mirror the focal length is negative and the image distance is negative. There will be a virtual image. The height of the image will be ¾ as tall as the object and it will be ¾ as far away from the mirror.

Solve: To solve this by ray tracing we draw the horizontal axis to scale and make the object an arbitrary height. First draw the ray that hits the center of the mirror, then trace it back behind the mirror with a dashed line.

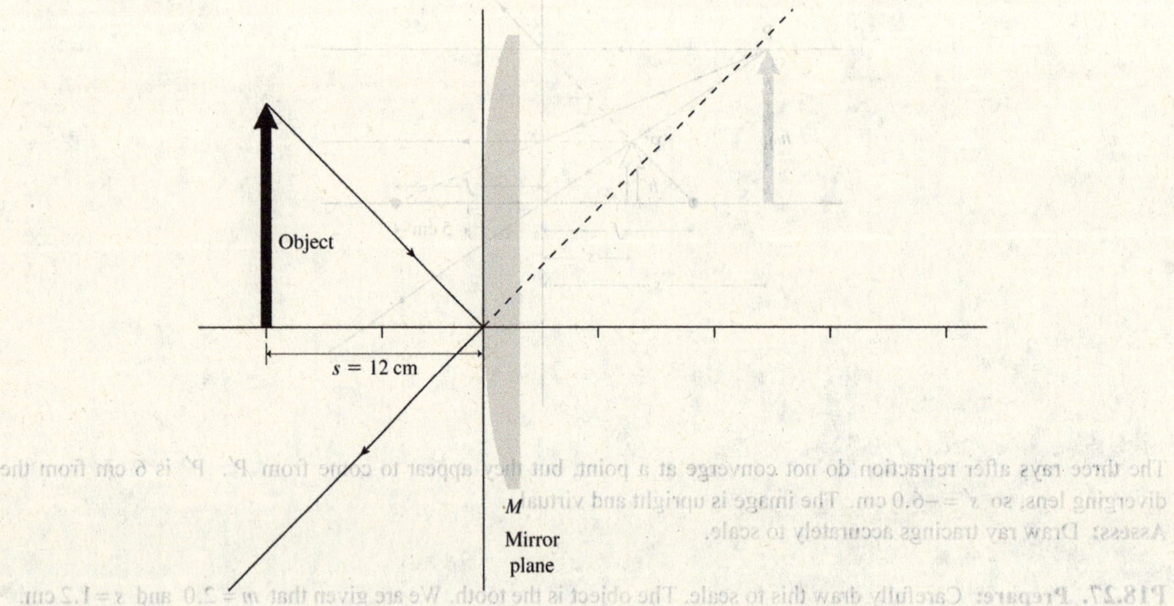

We know the top of the image must be on the dashed line. Use a ruler and see where on the dashed line it is ¾ as high as the object was tall. (This should also be ¾ the distance from the mirror as the object was, or 9.0 cm on the right of the mirror.)

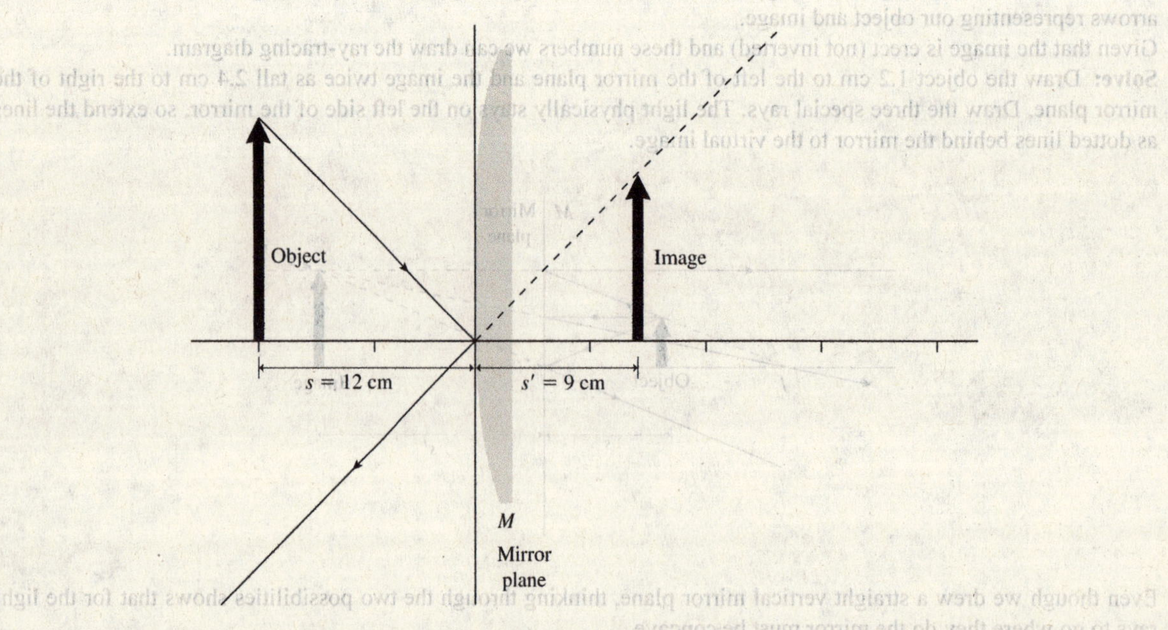

Now add another ray from the object, the one that first goes parallel to the optic axis. It bounces back as if coming from the focal point, so trace it back (right side of the mirror) with a dashed line to where the top of the image is, and then extend it back to the optic axis. This will demonstrate where the focal point is.

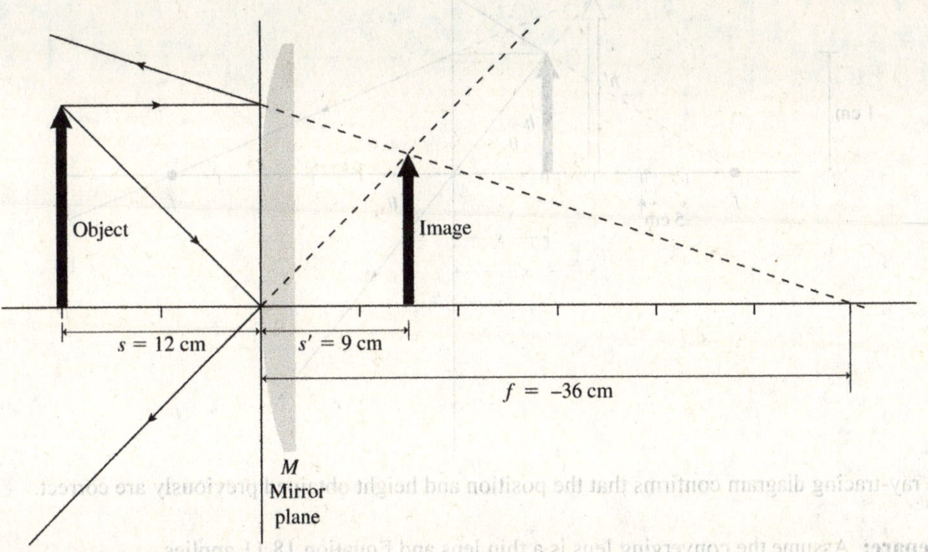

Use a ruler to measure the focal distance. It shows $f = -36$ cm.

Assess: We did not need the third ray that headed toward the focal point; furthermore, we didn't know where it was yet. This solution is independent of the height of the object; we could have drawn it any height as long as we made the image ¾ as tall.

This problem is much easier to do with the thin lens equation, but we get the same answer.

$$\frac{1}{s} + \frac{1}{s'} = \frac{1}{f} \Rightarrow \frac{1}{12\,\text{cm}} + \frac{1}{-9.0} = \frac{1}{f} \Rightarrow f = -36\,\text{cm}$$

P18.31. Prepare: Assume that the converging lens is a thin lens and Equation 18.11 applies.
Solve: Using the thin-lens formula,

$$\frac{1}{s} + \frac{1}{s'} = \frac{1}{f} \Rightarrow \frac{1}{10\,\text{cm}} + \frac{1}{s'} = \frac{1}{30\,\text{cm}} \Rightarrow \frac{1}{s'} = -\frac{1}{15\,\text{m}} \Rightarrow s' = -15\,\text{cm}$$

The image height is obtained from

$$m = -\frac{h'}{h} = -\frac{s'}{s} = -\frac{-15\,\text{cm}}{10\,\text{cm}} = +1.5$$

The image is upright and 1.5 times the object, that is, 1.5 cm high.
We also show in the next figure the ray-tracing diagram using the steps of Tactics Box 18.2. The three special rays that experience refraction do not converge at a point. Instead they appear to come from a point that is 15 cm on the same side as the object itself. Thus $s' = -15$ cm. The image is upright and has a height of $h' = 1.5$ cm.

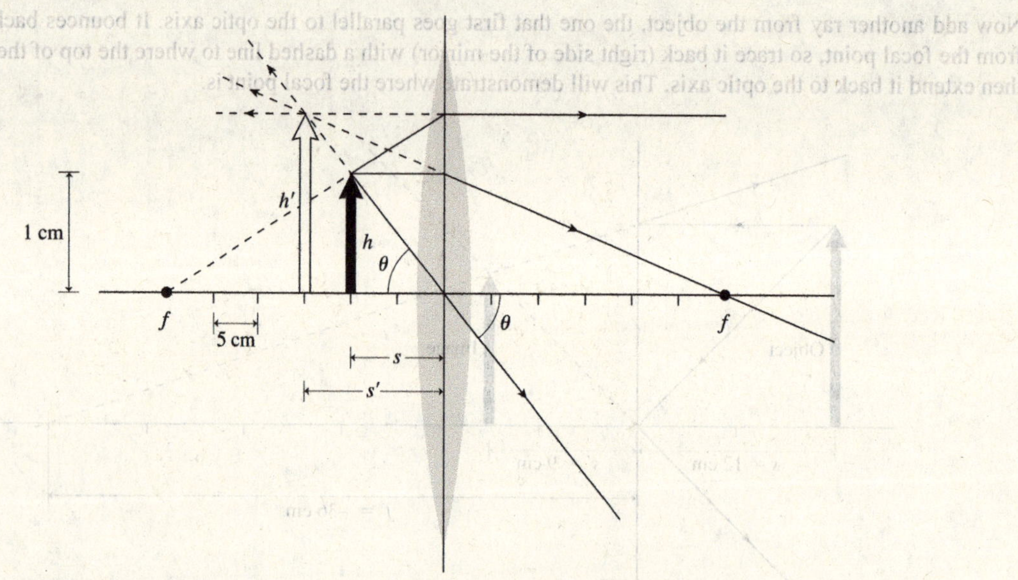

Assess: The ray-tracing diagram confirms that the position and height obtained previously are correct.

P18.33. Prepare: Assume the converging lens is a thin lens and Equation 18.11 applies.
Solve: Using the thin-lens formula,

$$\frac{1}{s}+\frac{1}{s'}=\frac{1}{f}\Rightarrow\frac{1}{75\text{ cm}}+\frac{1}{s'}=\frac{1}{30\text{ cm}}\Rightarrow\frac{1}{s'}=\frac{1}{50\text{ cm}}\Rightarrow s'=50\text{ cm}$$

The image height is obtained from

$$m=-\frac{s'}{s}=-\frac{50\text{ cm}}{75\text{ cm}}=-\frac{2}{3}$$

The image height is $h'=mh=(-2/3)(1\text{ cm})=-0.67\text{ cm}$. Because of the negative sign, the image is inverted.

We also show the ray-tracing diagram. After refraction, the three special rays converge and give an image 50 cm away from the converging lens. Thus, $s'=+50\text{ cm}$. The image is inverted and its height is 0.67 cm.

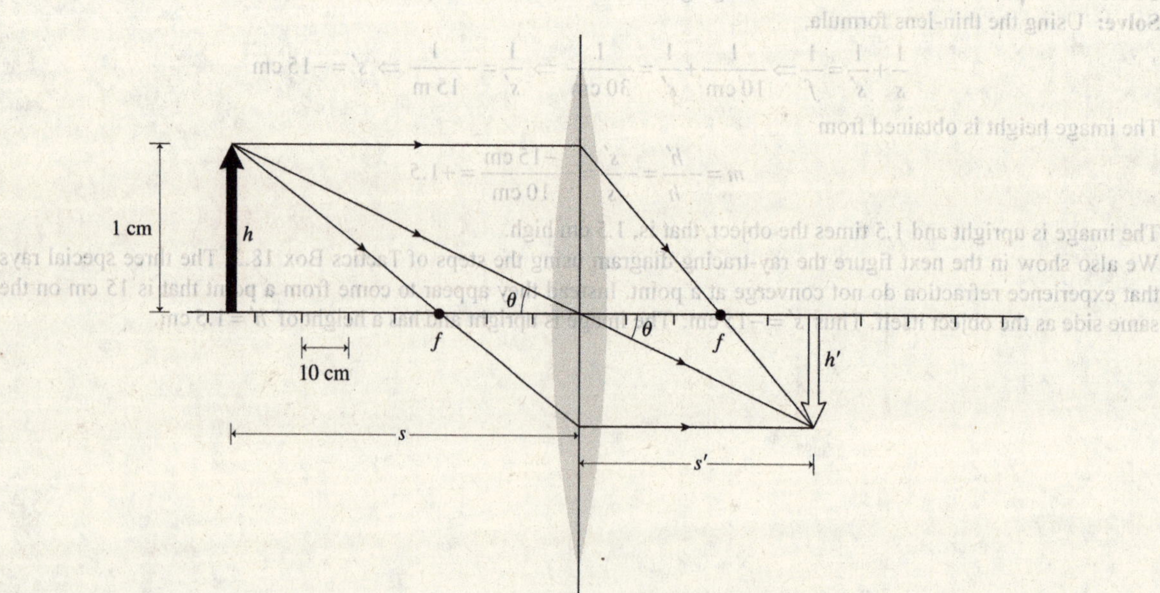

Assess: The ray-tracing diagram confirms that the position and height obtained previously are correct.

P18.37. Prepare: We'll first use the thin-lens equation to find the image position and then we'll use $h'/h = -s'/s$. We are given $s = 45\,\text{cm}$, $f = -25\,\text{cm}$, and $h = 3.0\,\text{cm}$.

Solve:

$$\frac{1}{s'} = \frac{1}{f} - \frac{1}{s} = \frac{1}{-25\,\text{cm}} - \frac{1}{45\,\text{cm}} = -0.0622\,\text{cm}^{-1}$$

So $s' = 1/-0.0622\,\text{cm}^{-1} = -16\,\text{cm}$. So the image is 16 cm behind the mirror.

$$h' = -h\frac{s'}{s} = -3.0\,\text{cm}\left(\frac{-16\,\text{cm}}{45\,\text{cm}}\right) = 1.1\,\text{cm}$$

Assess: A quick ray-tracing diagram confirms that both the position and height given are probably correct.

P18.39. Prepare: The object distance, image distance, and focal length are related by $1/s + 1/s' = 1/f$. The magnification is related to the object and image distance and height by $m = h'/h = -s'/s$.

Solve: The image position is obtained by

$$\frac{1}{s'} = \frac{1}{f} - \frac{1}{s} = \frac{1}{25\,\text{cm}} - \frac{1}{45\,\text{cm}} = \frac{+20}{1125\,\text{cm}}$$

or $s' = 56\,\text{cm}$.

The image height is obtained by

$$h' = -h(s'/s) = -(3.0\,\text{cm})(56\,\text{cm}/45\,\text{cm}) = -3.7\,\text{cm}$$

Assess: First note that the object is between the focal length and the radius of curvature of the concave mirror. This tells us that the image will be inverted (opposite orientation as the object), hence a negative image height; real (positive sign on the image distance); formed beyond the radius of curvature of the mirror; and have a magnification with a magnitude greater than one (the image appears to be taller than the object). The given information regarding images formed by a concave mirror is consistent with the calculated values.

P18.45. Prepare: To apply the law of reflection we need to know what direction the normal line points.

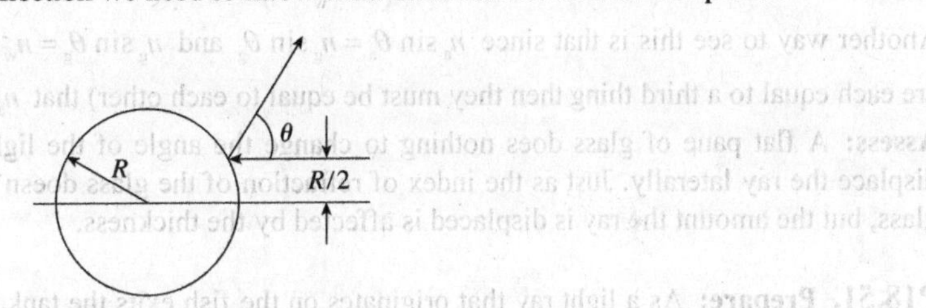

Solve: The right triangle tells us that the angle the normal makes with the horizontal is $\sin^{-1}(\text{opp}/\text{hyp}) = \sin^{-1}\frac{R/2}{R} = \sin^{-1}\left(\frac{1}{2}\right) = 30°$.

If the law of reflection is obeyed then $\theta = 2(30°) = 60°$.

Assess: The figure is drawn to scale and θ appears to be about 60°.

P18.47. Prepare: We will apply Snell's law at each interface and then come to an interesting conclusion (that we might expect).

$$n_1 \sin\theta_1 = n_2 \sin\theta_2$$

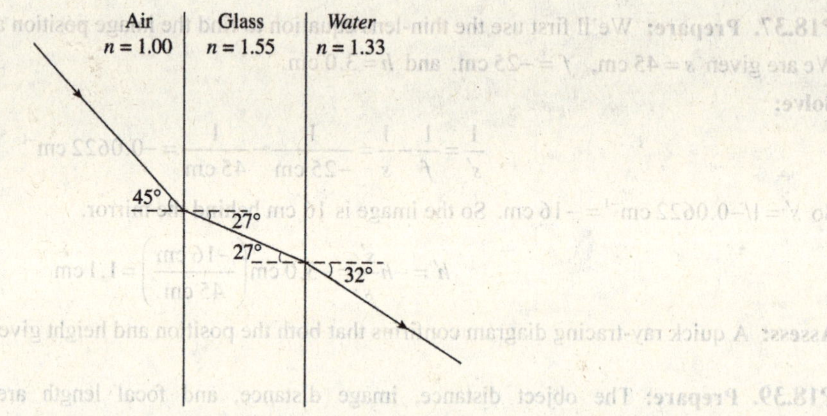

Solve: Use subscripts a, g, and w for air, glass, and water, respectively. Remember, all angles are measured from the normal.

(a)

$$\theta_g = \sin^{-1}\left(\frac{n_a \sin \theta_a}{n_g}\right) = \sin^{-1}\left(\frac{1.00 \sin 45°}{1.55}\right) = 27°$$

$$\theta_w = \sin^{-1}\left(\frac{n_g \sin \theta_g}{n_w}\right) = \sin^{-1}\left(\frac{1.55 \sin 27°}{1.33}\right) = 32°$$

So the laser enters the water at an angle of 32° from the normal.

(b) No, the result doesn't depend on the index of refraction of the pane of glass. Try the calculation directly from air to water to compare.

$$\theta_w = \sin^{-1}\left(\frac{n_a \sin \theta_a}{n_w}\right) = \sin^{-1}\left(\frac{1.00 \sin 45°}{1.33}\right) = 32°$$

Another way to see this is that since $n_a \sin \theta_a = n_g \sin \theta_g$ and $n_g \sin \theta_g = n_w \sin \theta_w$, then it must be true (if two things are each equal to a third thing then they must be equal to each other) that $n_a \sin \theta_a = n_w \sin \theta_w$.

Assess: A flat pane of glass does nothing to change the angle of the light entering the water, but it *does* slightly displace the ray laterally. Just as the index of refraction of the glass doesn't matter, neither does the thickness of the glass, but the amount the ray is displaced is affected by the thickness.

P18.51. Prepare: As a light ray that originates on the fish exits the tank, it will change directions and the angle of refraction in air will be greater than the angle of incidence in water.
Solve: The following sketch shows how the single object in the fish tank could have two images to an outside viewer.

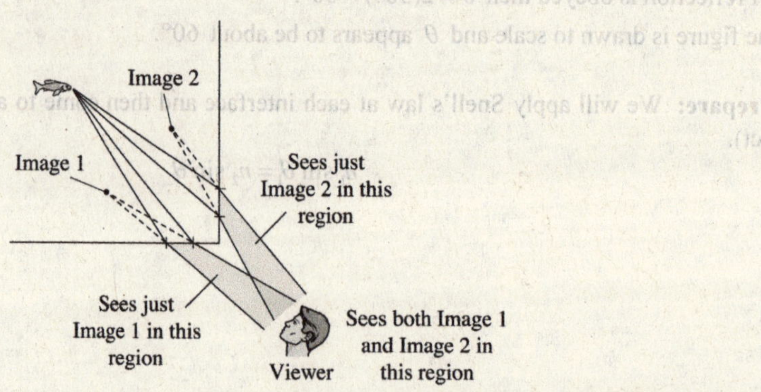

Assess: For clarity, the angles are greatly exaggerated—however the point that two images are possible is made.

P18.53. Prepare: Use the ray model of light. Assume that the target is a point source of light. We will apply Snell's law to the air–water boundary and use geometry in the following diagram.

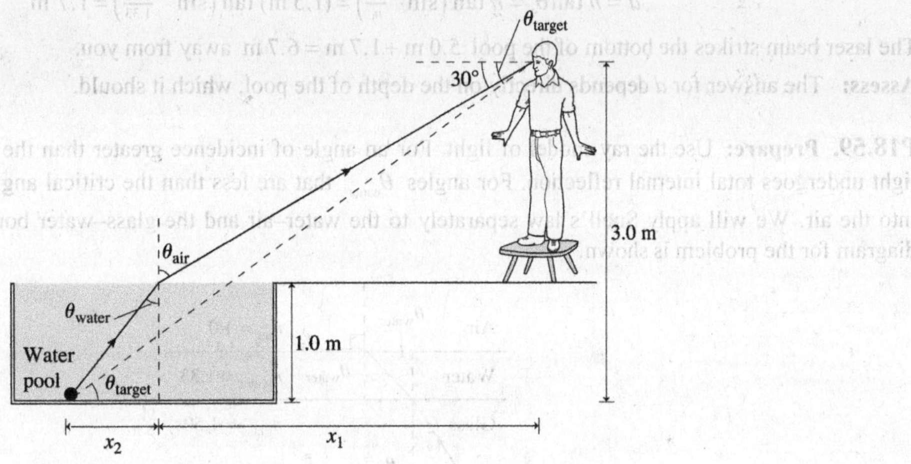

Solve: From the geometry of the figure with $\theta_{air} = 60°$,

$$\tan\theta_{air} = \frac{x_1}{2.0\ \text{m}} \Rightarrow x_1 = (2.0\ \text{m})(\tan 60°) = 3.464\ \text{m}$$

Let us find the horizontal distance x_2 by applying Snell's law to the air–water boundary. We have

$$n_{water}\sin\theta_{water} = n_{air}\sin\theta_{air} \Rightarrow \theta_{water} = \sin^{-1}\left(\frac{\sin 60°}{1.33}\right) = 40.63°$$

Using the geometry of the diagram,

$$\frac{x_2}{1.0\ \text{m}} = \tan\theta_{water} \Rightarrow x_2 = (1.0\ \text{m})\tan 40.63° = 0.858\ \text{m}$$

To determine θ_{target}, we note that

$$\tan\theta_{target} = \frac{3.0\ \text{m}}{x_1 + x_2} = \frac{3.0\ \text{m}}{3.464\ \text{m} + 0.858\ \text{m}} = 0.6941 \Rightarrow \theta_{target} = 35°$$

Assess: The previous ray-tracing diagram shown to scale would indicate θ_{target} to be slightly larger than 30°. An angle of 35° is reasonable.

P18.55. Prepare: Assume that for practical purposes the laser beam is parallel to the surface of the water. The beam will enter the water going in a direction θ_c from the normal. We want to know the total distance $(5.0\ \text{m}) + d$.

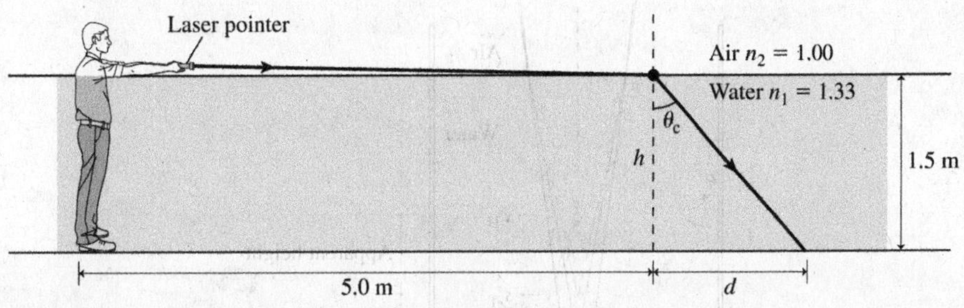

Solve: Note from the figure that $\tan\theta_c = \dfrac{d}{h}$.

$$d = h\tan\theta_c = h\tan\left(\sin^{-1}\tfrac{n_2}{n_1}\right) = (1.5\ \text{m})\tan\left(\sin^{-1}\tfrac{1.00}{1.33}\right) = 1.7\ \text{m}$$

The laser beam strikes the bottom of the pool $5.0\ \text{m} + 1.7\ \text{m} = 6.7\ \text{m}$ away from you.

Assess: The answer for d depends directly on the depth of the pool, which it should.

P18.59. Prepare: Use the ray model of light. For an angle of incidence greater than the critical angle, the ray of light undergoes total internal reflection. For angles θ_{water} that are less than the critical angle, light will be refracted into the air. We will apply Snell's law separately to the water–air and the glass–water boundaries. The ray-tracing diagram for the problem is shown.

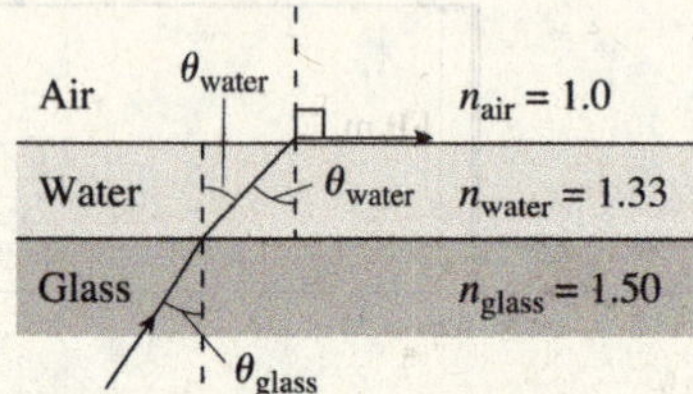

Solve: Snell's law at the water–air boundary is $n_{\text{air}}\sin\theta_{\text{air}} = n_{\text{water}}\sin\theta_{\text{water}}$. Because the maximum angle of θ_{air} is 90°, we have

$$(1.0)\sin 90° = 1.33\sin\theta_{\text{water}} \Rightarrow \theta_{\text{water}} = \sin^{-1}\left(\frac{1}{1.33}\right) = 48.75°$$

Applying Snell's law again to the glass–water boundary,

$$n_{\text{glass}}\sin\theta_{\text{glass}} = n_{\text{water}}\sin\theta_{\text{water}} \Rightarrow \theta_{\text{glass}} = \sin^{-1}\left(\frac{n_{\text{water}}}{n_{\text{glass}}}\sin\theta_{\text{water}}\right) = \sin^{-1}\left(\frac{1.33(\sin 48.75°)}{1.50}\right) = 42°$$

Thus 42° is the maximum angle of incidence onto the glass for which the ray emerges into the air.

Assess: The index of refraction of the water canceled out, and the thickness of the water was irrelevant.

P18.61. Prepare: Represent the diver's head and toes as point sources. Use the ray model of light. Paraxial rays from the head and the toes of the diver refract into the air and then enter into your eyes. When these refracted rays are extended into the water, the head and the toes appear elevated toward you.

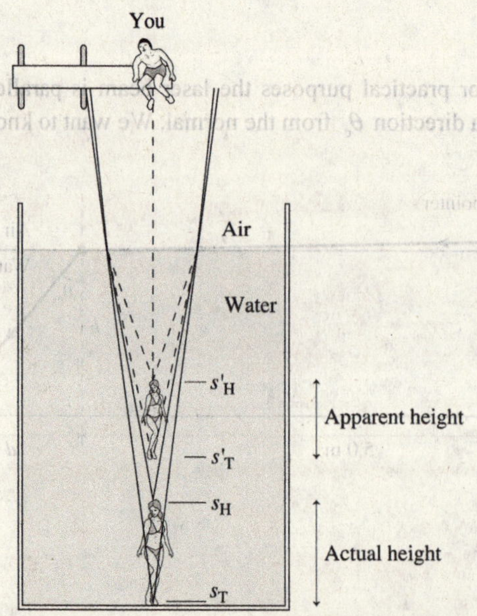

Solve: Using Equation 18.7,

$$s'_T = \frac{n_2}{n_1} s_T = \frac{n_{air}}{n_{water}} s_T \quad \text{and} \quad s'_H = \frac{n_{air}}{n_{water}} s_H$$

Subtracting the two equations, her apparent height is

$$s'_H - s'_T = \frac{n_{air}}{n_{water}} (s_H - s_T) = \frac{1.0}{1.33} (150 \text{ cm}) = 110 \text{ cm}$$

Assess: The previous ray-tracing diagram shows that the diver should appear smaller. The value of 110 cm is reasonable.

P18.63. Prepare: The microscope slide makes the object (amoeba) appear closer than it actually is. Knowing the object distance (the thickness of the slide) and the index of refraction of air and glass, we can determine the distance the microscope objective must be moved.

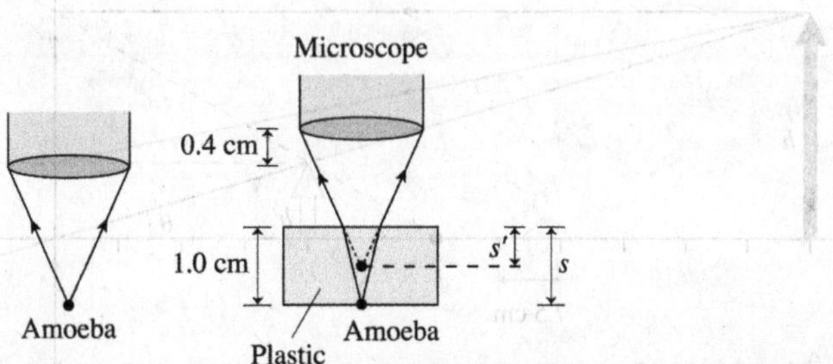

Solve: The apparent position of the amoeba below the surface of the glass microscope slide is

$$s' = s(n_A/n_G) = 0.15 \text{ mm} (1.0/1.5) = 0.10 \text{ mm}$$

The amoeba appears to be 0.05 mm closer than it really is, so the microscope objective must be moved 0.05 mm in order to bring it back in focus. The amoeba was in sharp focus; you in essence brought it closer to the objective and it was not in focus. In order to get in focus again, you need to increase the distance between the object (the apparent amoeba) and the objective. This can be done by either lowering the viewing table or raising the objective.

Assess: It will be interesting to learn more about optical instruments (including microscopes) in the next chapter.

P18.69. Prepare: We will use ray tracing to locate the image. Assume that the converging lens is a thin lens.
Solve:

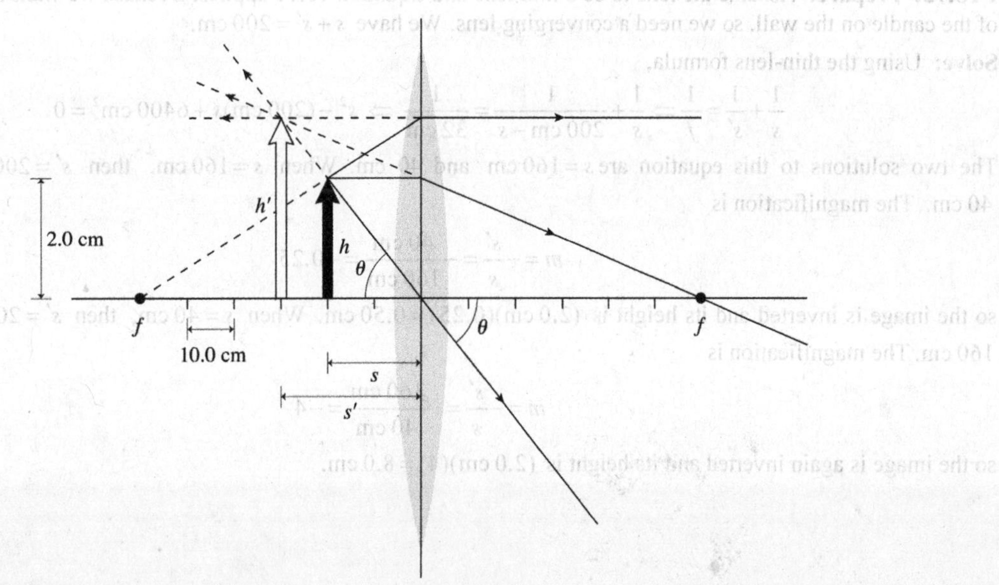

The figure shows the ray-tracing diagram using the steps of Tactics Box 18.2. The three special rays that experience refraction do not converge at a point. Instead they appear to come from a point that is 30 cm on the same side as the object itself. Thus $s' = -30$ cm. The image is upright and from Equation 18.9 has a height of $h' = 3.0$ cm.

P18.71. Prepare: We will use ray tracing according to Tactics Box 18.3 to locate the image. Assume the diverging lens is a thin lens.
Solve:

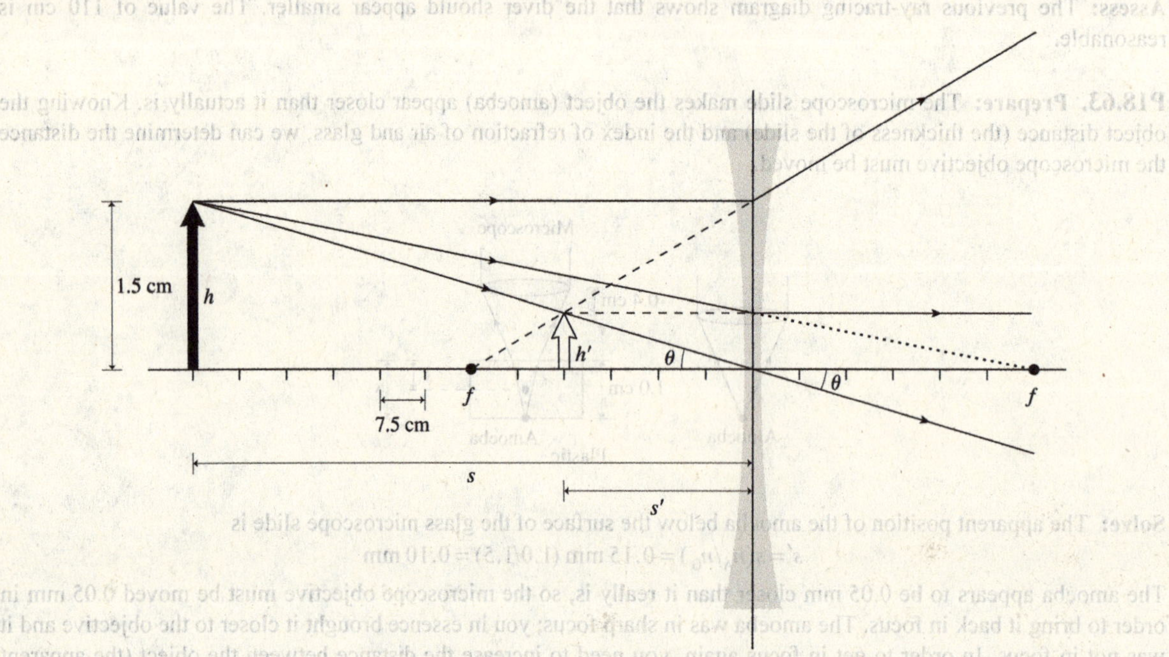

The figure shows the ray-tracing diagram using the steps of Tactics Box 18.3. After refraction from the diverging lens, the three special rays do not converge. However, the rays appear to meet at a point that is 30 cm on the same side as the object. So $s' = -30$ cm. The image is upright and using Equation 18.9 has a height of 0.50 cm.

P18.73. Prepare: Assume the lens to be a thin lens and Equation 18.11 applies. Because we want to form an image of the candle on the wall, so we need a converging lens. We have $s + s' = 200$ cm.
Solve: Using the thin-lens formula,

$$\frac{1}{s} + \frac{1}{s'} = \frac{1}{f} \Rightarrow \frac{1}{s} + \frac{1}{200\text{ cm} - s} = \frac{1}{32\text{ cm}} \Rightarrow s^2 - (200\text{ cm})s + 6400\text{ cm}^2 = 0$$

The two solutions to this equation are $s = 160$ cm and 40 cm. When $s = 160$ cm, then $s' = 200$ cm $- 160$ cm $= 40$ cm. The magnification is

$$m = -\frac{s'}{s} = -\frac{40\text{ cm}}{160\text{ cm}} = -0.25$$

so the image is inverted and its height is $(2.0\text{ cm})(0.25) = 0.50$ cm. When $s = 40$ cm, then $s' = 200$ cm $- 40$ cm $= 160$ cm. The magnification is

$$m = -\frac{s'}{s} = -\frac{160\text{ cm}}{40\text{ cm}} = -4$$

so the image is again inverted and its height is $(2.0\text{ cm})(4) = 8.0$ cm.

P18.75. Prepare: The object distance, image distance, and focal length are related by $1/s + 1/s' = 1/f$.

Solve: The end (let's agree to call it the near end—ne) of the meter stick closest to the mirror has its image located at

$$1/s'_{ne} = 1/f - 1/s_{ne} = 1/40\,\text{cm} - 1/60\,\text{cm} = 1/120\,\text{cm} \quad \text{or} \quad s'_{ne} = 120\,\text{cm}$$

The end (let's agree to call it the far end—fe) of the meter stick farthest from the mirror has its image located at

$$1/s'_{fe} = 1/f - 1/s_{fe} = 1/40\,\text{cm} - 1/160\,\text{cm} = 3/160\,\text{cm} \quad \text{or} \quad s'_{fe} = 53.3\,\text{cm}$$

The image of the meter stick appears to be $120\,\text{cm} - 53.3\,\text{cm} = 67\,\text{cm}$.

Assess: It's interesting to note that the image of the end of the meter stick closest to the mirror appears to be farther away when viewed in the mirror.

OPTICAL INSTRUMENTS

Q19.1. Reason: If the viewing screen is fairly close to a hole or gap in a barrier, then the spot of light on the viewing screen will have the shape of the gap or hole. But if the viewing screen (in this case the ground) is far away, then the gap acts like a pinhole camera and the image on the screen looks like an inverted image of the object. The illuminated spots are circles because the sun is round.

Assess: The spots may not appear as extremely sharp circles because of diffraction and the size of the holes. In a strict ray model with a very small pinhole the image would be sharp.

When I happened to be visiting San Antonio during a recent partial solar eclipse, I noticed that the spots of light on the ground made by sunlight coming through the leaves were crescent-shaped. I also had two cards, one with a pinhole and one as the screen, with me to view the eclipse. The image on the card was sharper than the spots on the ground because the pinhole was smaller than the gaps between the leaves. Of course, the smaller the pinhole, the dimmer the image because less light gets through.

Q19.5. Reason: The object distance, image distance, and focal length are related by

$$\frac{1}{s} + \frac{1}{s'} = \frac{1}{f}$$

Inspecting this expression, we see that for a fixed focal length, as the object distance gets smaller the image distance must get larger in order to maintain the equality. Due to the construction of the camera, there is a limit to how big the image distance can be and still get a sharp image on the detectpr. In order to allow smaller object distances (which is the case for a close-up shot), the photographer can use a smaller focal length lens.

Assess: If the focal length gets smaller, the object distance can decrease and the image distance still fits the camera.

Q19.7. Reason: I moved my finger to the left and followed it with my right eye until the cross disappeared. So the image of the cross was falling on my right retina on its left side.

Assess: The blind spots are "located nasally" (toward the nose) on the retina and so they are away from the nose in the visual field.

Q19.11. Reason: The resolving power of a microscope is given by $\mathrm{RP} = 0.61\lambda_0 / \mathrm{NA}$. The numerical aperture is a property of the microscope's *objective* lens, and is not related to any properties of the eyepiece. If you switch eyepieces, the resolving power will not be affected.

Assess: If your friend had switched objective lenses, the resolving power of the microscope could have been improved.

Q19.15. Reason: The card is red because it reflects red light and absorbs the other colors. When it is illuminated by red light, the red light reflects off the card into your eyes and you see the red card as red.

If the card is illuminated with blue light the light is all absorbed. No light is reflected, so the card looks black.

Assess: If you illuminate the card with white light and look at it through a blue filter it will again look black because the red light reflected by the card is not passed by the blue filter.

Q19.17. Reason: The magnification of the microscope is given by
$$M = m_o M_e = -(L/f_o)(25\text{ cm}/f_e) = -(20\text{ cm})(25\text{ cm})/(f_o f_e) = -500\text{ cm}^2/(f_o f_e)$$
Or
$$f_o f_e = -500\text{ cm}^2/M = -500\text{ cm}^2/100 = -5.0\text{ cm}^2$$
A combination of focal lengths that would allow this is $f_o = 1\text{ cm}, f_e = 5\text{ cm}$.
The correct choice is C.
Assess: These are reasonable values for a microscope.

Q19.21. Reason: In this case we would like to create a virtual image with an image distance of $s' = -100\text{ cm}$ for
an object with an object distance of $s = 40\text{ cm}$. This can be done with a lens of focal length
$$\frac{1}{f} = \frac{1}{s} + \frac{1}{s'} = \frac{1}{40\text{ cm}} - \frac{1}{100\text{ cm}} = \frac{3}{200\text{ cm}} = 0.015\text{cm}^{-1} \quad \text{or} \quad f = 66.7\text{ cm}$$
The refractive power of the desired lens is
$$P = \frac{1}{f} = \frac{1}{0.667\text{ cm}} = +1.5\text{ D}$$
The correct choice is C.
Assess: A converging lens is needed to correct the sight of a farsighted person.

Q19.23. Reason: A red object reflects only red and absorbs all other colors, a white object reflects all colors and a
blue object reflects only blue and absorbs all other colors. As a result, if red light shines on a red cup, a white card,
and a blue toy they will respectively appear to be red, red, and black. The correct choice is C.
Assess: The primary additive colors reflect only that color and white reflects all colors.

Problems

P19.3. Prepare: Knowing that the vertical angles are equal we can use the tangent trigonometry function to
determine half of the height and then the height.

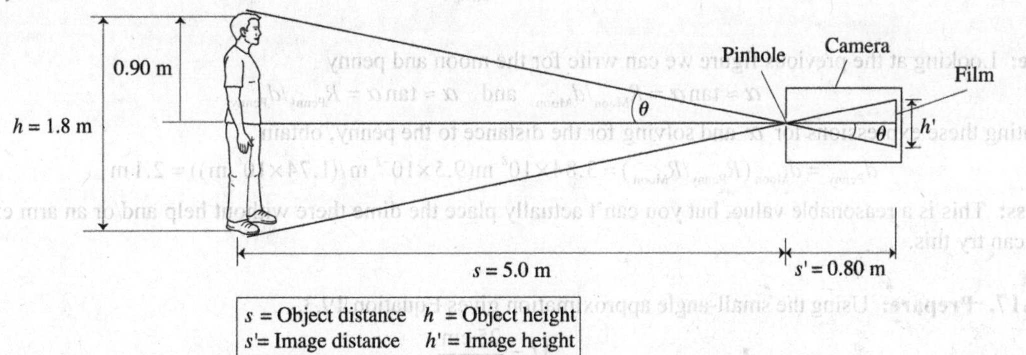

Solve: Using the tangent function we can write $\tan\theta = (h/2)/s = (h'/2)/s'$.
Solving for h' obtain $h' = h(s'/s) = 1.8\text{ m}(0.80\text{ m}/(5.0\text{ m})) = 0.29\text{ m}$.
Assess: The box must be at least 0.29 m wide. Given that it is 0.80 m long, this is not an unreasonable size for
a box.

P19.5. Prepare: Assume the bridge is quite far away so we can approximate $s = \infty$.
Solve: The ratio of the focal lengths must be the same as the ratio of the widths of the images.
$$\frac{f_2}{f_1} = \frac{s_2'}{s_1'}\left(\frac{s + s_1'}{s + s_2'}\right) \approx \frac{h_2'}{h_1'} \Rightarrow f_2 = f_1\frac{h_2'}{h_1'} = (50\text{ mm})\left(\frac{12\text{ mm}}{36\text{ mm}}\right) = 17\text{ mm}$$
Assess: The field of view is inversely proportional to the focal length.

P19.9. Prepare: Since the woman has a far point of 3.0 m, she cannot see objects clearly that are more than 3.0 m away. We need a lens that will make distant objects $(s = \infty)$ appear to be 3.0 m away on the same side of the lens $(s' = -3.0 \text{ m})$. We can use the lens equation to determine $1/f$ and hence P since $P = 1/f$.

Solve: The power of the desired lens is
$$P = 1/f = 1/s + 1/s' = 1/\infty - 1/(3.0 \text{ m}) = -1/(3.0 \text{ m}) = -0.33 \text{ D}$$

A negative power for a lens indicates a diverging lens.

Assess: This is a small power compared to Example 19.3 in the text. However, the far point for the woman is almost 10 times the far point of the person in the example.

P19.11. Prepare: Mary can focus only on distant objects $(s' = \infty)$. The book will be held 0.50 m away $(s = 0.5 \text{ m})$. We can use the lens equation to determine $1/f$ and hence P since $P = 1/f$.

Solve: The power of the desired lens is
$$P = 1/f = 1/s + 1/s' = 1/(0.50 \text{ m}) + 1/\infty = 1/(0.50 \text{ m}) = 2.0 \text{ D}$$

A positive power for a lens indicates a converging lens.

Assess: This is a moderately strong lens, but reading glasses like this are readily available.

P19.15. Prepare: For this case we want two objects which vary greatly in size to project the same size image on your retina. This is exactly like the case shown in Figure 19.9 (b). So let's start by drawing a figure similar to the one shown in Figure 19.9 (b).

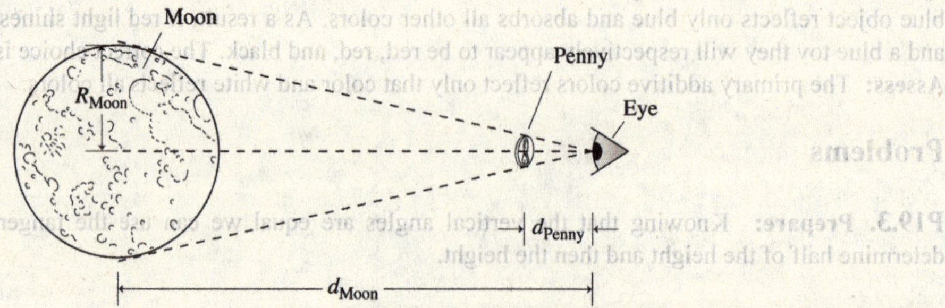

Solve: Looking at the previous figure we can write for the moon and penny
$$\alpha \approx \tan\alpha = R_{\text{Moon}}/d_{\text{Moon}} \quad \text{and} \quad \alpha \approx \tan\alpha = R_{\text{Penny}}/d_{\text{Penny}}$$

Equating these expressions for α and solving for the distance to the penny, obtain
$$d_{\text{Penny}} = d_{\text{Moon}}(R_{\text{Penny}}/R_{\text{Moon}}) = 3.84 \times 10^8 \text{ m}(9.5 \times 10^{-3} \text{ m}/(1.74 \times 10^6 \text{ m})) = 2.1 \text{ m}$$

Assess: This is a reasonable value, but you can't actually place the dime there without help and/or an arm extension. You can try this.

P19.17. Prepare: Using the small-angle approximation gives Equation 19.3.
$$M = \frac{25 \text{ cm}}{f}$$

The magnification $M = 5$ is positive because the image is upright.

Solve:
$$f = \frac{25 \text{ cm}}{M} = \frac{25 \text{ cm}}{5} = 5.0 \text{ cm}$$

The object should be placed almost the same distance, 5.0 cm, away from the lens.

Assess: These numbers all seem realistic.

P19.19. Prepare: Equation 19.4 relates the length of the microscope tube, the magnification, and the focal length of the objective.

$$M_o = -\frac{L}{f_o}$$

Equation 19.5 gives the total magnification.

$$M = M_o M_e = -\frac{L}{f_o} \frac{25\ \text{cm}}{f_e}$$

Solve: **(a)** Solve Equation 19.4 for f_o. We use a negative magnification because the image is inverted.

$$f_o = -\frac{L}{M_o} = -\frac{12.0\ \text{cm}}{-10} = 1.20\ \text{cm}$$

(b) Solve Equation 19.5 for f_e. We use a negative magnification because the image is inverted.

$$f_e = -\frac{L}{f_o} \frac{25\ \text{cm}}{M} = -\frac{12.0\ \text{cm}}{1.20\ \text{cm}} \frac{25\ \text{cm}}{(-150)} = 1.67\ \text{cm}$$

Assess: These numbers are all within realistic ranges.

P19.21. Prepare: Example 19.6 shows how to find the focal length of a microscope objective.
The "15×" means that the objective has a magnification M_o of -15. We can use Equation 19.4, taking the tube length L as 160 mm, which we've seen is the standard length for a biological microscope.
The object is placed very near the focal plane of the objective, that is, $s \approx f_o$.

Solve:

$$s \approx f_o = -\frac{L}{M_o} = -\frac{160\ \text{mm}}{-15} = 10.67\ \text{mm} \approx 11\ \text{mm}$$

Assess: The result is a little larger than the result in Example 19.6, just as we expect because our magnification was a little less. The magnification of the eyepiece was unneeded information.

P19.25. Prepare: We'll use the thin lens equation twice. The first time will give us the image from the converging lens, and that image will then be used as the object for the diverging lens.

Solve: For the converging lens, $f = 5.0\ \text{cm}$ and $s = 4.0\ \text{cm}$.

$$s' = \frac{1}{\frac{1}{f} - \frac{1}{s}} = \frac{fs}{s - f} = \frac{(5.0\ \text{cm})(4.0\ \text{cm})}{4.0\ \text{cm} - 5.0\ \text{cm}} = -20\ \text{cm}$$

That s' is negative means it is a virtual image on the same side of the lens as the object.
The magnification is

$$M = -\frac{s'}{s} = -\frac{-20\ \text{cm}}{4.0\ \text{cm}} = 5.0$$

The magnification is positive because the virtual image is upright. We now use the magnification to compute h'.

$$h' = Mh = 5.0(1.0\ \text{cm}) = 5.0\ \text{cm}$$

To summarize so far, we have an image from the first lens that is upright, 20 cm to the left of the lens, and 5.0 cm tall.
That image now becomes the object for the diverging lens on the right. For the diverging lens $f = -8.0\ \text{cm}$ and $s = 20\ \text{cm} + 12\ \text{cm} = 32\ \text{cm}$.

$$s' = \frac{1}{\frac{1}{f} - \frac{1}{s}} = \frac{fs}{s - f} = \frac{(-8.0\ \text{cm})(32.0\ \text{cm})}{32.0\ \text{cm} - (-8.0\ \text{cm})} = -6.4\ \text{cm}$$

That s' is negative means it is a virtual image on the same side of the lens as the object.

The magnification is

$$M = -\frac{s'}{s} = -\frac{-6.4\,\text{cm}}{32\,\text{cm}} = 0.20$$

The magnification is positive because the virtual image is upright. We now use the magnification to compute h'.

$$h' = Mh = 0.20(5.0\,\text{cm}) = 1.0\,\text{cm}$$

The image is upright (because h' is positive).

At the end, the final image is virtual, upright, 6.4 cm to the left of the diverging lens (almost halfway between the two lenses, or 9.6 cm from the object), and 1.0 cm tall.

Assess: The final image is the same height as the original object because the product of the two magnifications is 1.0. The result of this problem could be verified with ray diagrams or real-life lenses.

P19.27. Prepare: The refractive power is the inverse of the focal length. First convert ft to m: $f_o = 57\,\text{ft} = 17.4\,\text{m}$. Equation 19.6 gives the magnification of a telescope in terms of the focal lengths of the two lenses.

$$M = -\frac{f_o}{f_e}$$

Solve: (a)

$$P = \frac{1}{f} = \frac{1}{57\,\text{ft}} = \frac{1}{17.4\,\text{m}} = 0.0576\,\text{m}^{-1} = 0.0576\,\text{D} \approx 0.058\,\text{D}$$

to two significant figures.

(b) The "$1000\times$" means the magnification is -1000 since the image is inverted. Solve Equation 19.6 for f_e.

$$f_e = -\frac{f_o}{M} = -\frac{17.4\,\text{m}}{-1000} = 17.4\,\text{mm} \approx 17\,\text{mm}$$

Assess: The refractive power of the objective is quite small, as is expected for a long focal length. The eyepiece, however, has a smaller focal length, and 17 mm is in a reasonable range.

P19.31. Prepare: Table 19.1 tells us the wavelength of the deepest red is about 700 nm and the wavelength of the deepest violet is about 400 nm. To compute the angle of refraction we need to know the index of refraction for flint glass for the different colors; this dispersion curve appears in Figure 19.19. Reading from the graph, $n_{\text{red}} \approx 1.57$ and $n_{\text{violet}} \approx 1.60$. The strategy now is to compute θ_2 for each color from Snell's law (Equation 18.2): $n_1 \sin\theta_1 = n_2 \sin\theta_2$ and then use trigonometry to find the exit location. Note that the angle in Figure P19.31 (60°) is the complement of θ_1 in Snell's law (which is measured from the normal); i.e., $\theta_1 = 30°$. Assume that the ray of light enters the glass from the air so that $n_1 = n_{\text{air}} = 1.00$.

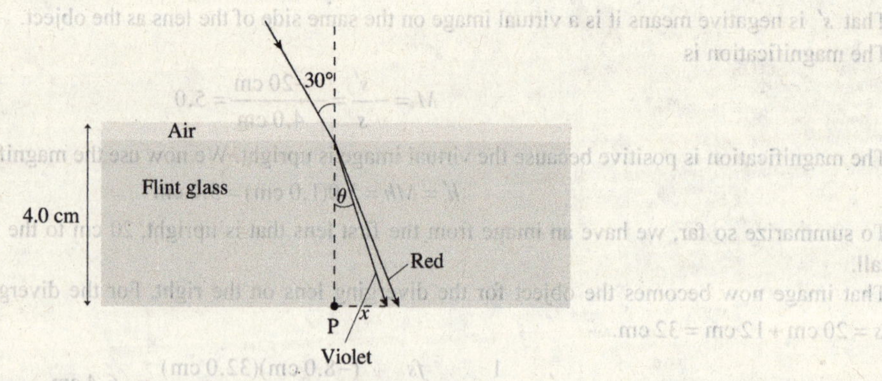

Look at the right triangle formed by the ray in the glass, the normal line, and the bottom edge of the glass where the ray exits. Convince yourself that the distance from point P to where the ray exits is $x = (4.0\,\text{cm})\tan\theta_2$. We now have all the preliminary pieces needed to solve this problem.

Solve: Solve Snell's law for θ_2 for each color.

$$(\theta_2)_{\text{red}} = \sin^{-1}\left(\frac{n_{\text{air}} \sin \theta_1}{n_{\text{red}}}\right) = \sin^{-1}\left(\frac{1.00 \sin 30°}{1.57}\right) = 18.57°$$

$$(\theta_2)_{\text{violet}} = \sin^{-1}\left(\frac{n_{\text{air}} \sin \theta_1}{n_{\text{violet}}}\right) = \sin^{-1}\left(\frac{1.00 \sin 30°}{1.60}\right) = 18.21°$$

Now compute the distance x from point P of each exiting ray.

$$x_{\text{red}} = (4.0 \text{ cm}) \tan (\theta_2)_{\text{red}} = (4.0 \text{ cm}) \tan (18.57°) = 1.344 \text{ cm}$$

$$x_{\text{violet}} = (4.0 \text{ cm}) \tan (\theta_2)_{\text{violet}} = (4.0 \text{ cm}) \tan (18.21°) = 1.316 \text{ cm}$$

The two rays exit the glass a distance apart that is the difference of these two distances. $\Delta x = 1.344 \text{ cm} - 1.316 \text{ cm} = 0.028 \text{ cm} = 0.28 \text{ mm}$.

The index of refraction indicates how much the light slows down in the medium, and consequently how much it refracts. The index of refraction in flint glass for violet light is higher than the index for red light, so the violet light is bent more and exits closer to point P.

Assess: The angles of the two rays are different, but not by much, so the rays aren't very far apart after traversing a 4 cm piece of glass. Generally the difference in index of refraction is greater for different materials than it is for different colors within a given material. We had to make our intermediate calculations to four significant figures to get two significant figures on the final answer. This is tenuous because it is difficult to estimate n from the graph that accurately.

P19.35. Prepare: The resolving power is $\text{RP} = d_{\text{min}} = 0.61\lambda_0 / \text{NA}$, where the numerical aperture is $\text{NA} = \sin \phi_0$.

Solve: The require numerical aperture is

$$\text{NA} = \sin \phi_0 = \frac{0.61\lambda_0}{d_{\text{min}}} = \frac{0.61(550 \text{ nm})}{400 \text{ nm}} = 0.839$$

so that $\phi_0 = \sin^{-1}(0.839) = 57.0°$.

From Figure 19.27, you can see that

$$\tan \phi_0 = \frac{D/2}{f} \Rightarrow D = 2f \tan \phi_0 = 2(1.6 \text{ mm}) \tan (57.0°) = 4.9 \text{ mm}$$

Assess: This seems reasonable for a lens with a focal length of 1.6 mm.

P19.37. Prepare: The object distance, image distance, and focal length of a lens are related by $1/s + 1/s' = 1/f$. The magnification is related to the object and image distance and height by $m = h'/h = -s'/s$. In this problem the magnifier is not used in the usual way, i.e., the object is not placed at the focal point. The angular size of an object when viewed at the near point is $\theta_\text{o} = h/d_{\text{near}}$; the angular size of that object when viewed through the magnifying lens is $\theta = h'/s'$, and the angular magnification of the lens is

$$M = \frac{\theta}{\theta_\text{o}} = \frac{h'/s'}{h/(25 \text{ cm})} = \frac{h'}{h}\frac{(25 \text{ cm})}{s'}$$

Solve: We can solve the lens equation for the image distance as follows:

$$1/s' = 1'/f - 1/s = (s - f)/(sf) \quad \text{or} \quad s' = \frac{sf}{s - f} = \frac{(8.0 \text{ cm})(10.0 \text{ cm})}{-2.0 \text{ cm}} = -40 \text{ cm}$$

The magnification is

$$m = -s'/s = h'/h = -(-40 \text{ cm})/(8 \text{ cm}) = 5$$

Finally, the angular magnification of the lens is

$$M = \left(\frac{h'}{h}\right)\frac{25 \text{ cm}}{s'} = (5)\frac{25 \text{ cm}}{-40 \text{ cm}} = 3.1$$

Assess: If Jason used the magnifier in the usual way, with the object at the focus, its magnification would be $M = (25 \text{ cm})/f = 2.5$, so our result seems reasonable.

P19.41. Prepare: Each lens is a thin lens. The image of the first lens is the object for the second lens.
Solve: **(a)** The following figure shows the two lenses and a ray-tracing diagram. From the ray-tracing diagram, we find that the image is 20 cm in front of the second lens and the height of the final image is 2.0 cm. The ray-tracing shows that the lens combination will produce a virtual, inverted image in front of the second lens.

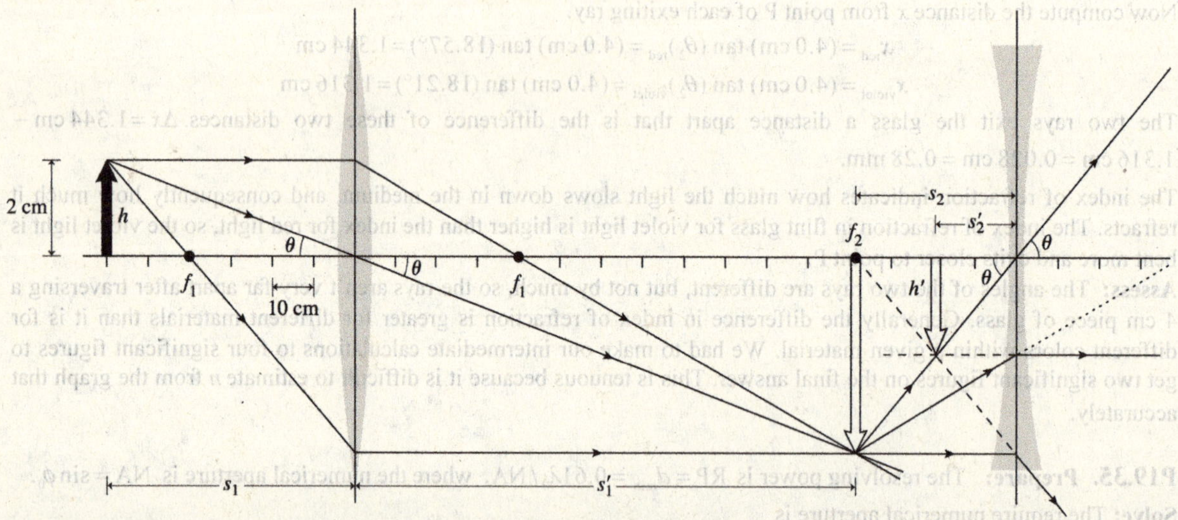

From the ray diagram we see that the final image is 20 cm in front of the second lens, and its height is -2.0 cm. The minus sign informs us that the image is inverted relative to the initial object and that it is virtual.
(b) $s_1 = 60$ cm is the object distance of the first lens. Its image, which is a real image, is found from the thin-lens equation

$$\frac{1}{s_1'} = \frac{1}{f_s} - \frac{1}{s_1} = \frac{1}{40 \text{ cm}} - \frac{1}{60 \text{ cm}} = \frac{1}{120 \text{ cm}} \Rightarrow s_1' = 120 \text{ cm}$$

The magnification of the first lens is

$$m_1 = -\frac{s_1'}{s_1} = -\frac{120 \text{ cm}}{60 \text{ cm}} = -2$$

The image of the first lens is now the object for the second lens. The object distance is $s_2 = 160 \text{ cm} - 120 \text{ cm} = 40 \text{ cm}$. A second application of the thin-lens equation yields the following

$$\frac{1}{s_2'} = -\frac{1}{s_2} + \frac{1}{f_2} = \frac{-1}{+40 \text{ cm}} + \frac{1}{-40 \text{ cm}} \Rightarrow s_2' = -20 \text{ cm}$$

The magnification of the second lens is

$$m_2 = -\frac{s_2'}{s_2} = -\frac{-20 \text{ cm}}{40 \text{ cm}} = +0.5$$

The overall magnification is $M = m_1 m_2 = (-2)(0.5) = -1.0$. The height of the final image is $(+1.0)(2.0 \text{ cm}) = 2.0$ cm. The image is inverted because M has a negative sign.

Assess: The calculated values in part **(b)** are in agreement with those found in part **(a)**.

P19.43. Prepare: The size of an image is related to the size of the object and magnification by $h' = Mh$.
Solve: The size of the real image of the specimen created by the objective lens is

$$h' = Mh = 5(1.0 \text{ mm}) = 5.0 \text{ mm}$$

Assess: If the lens provides a magnification of 5×, then the image of a 1 mm object will be 5 mm.

P19.45. Prepare: The total angular magnification of a microscope is given by $M = -L(25\,\text{cm})/(f_o f_e)$.

Solve: Notice that the quantity $M/L = -25\,\text{cm}/(f_o f_e)$ is a constant. Using "primes" to indicate the new situation we have $M'/L' = -25\,\text{cm}/(f_o f_e)$. Equating these two situations and solving for M' obtain the following:

$$M' = M(L'/L) = 200(200\,\text{mm}/(160\,\text{mm})) = 250$$

Assess: We expected a greater magnification from a microscope with a greater tube length.

P19.49. Prepare: Since Martha has a far point of 35 cm, she cannot see objects clearly that are more than 35 cm away. We need a lens that will make distant objects ($s = \infty$) appear to be 35 cm away on the same side of the lens ($s' = -35$ cm). We can use the lens equation to determine the focal length of her glasses, then use that to find the closest object on which she can focus.

Solve: The power of the desired lens is

$$1/f = 1/s + 1/s' = 1/\infty - 1/(35\,\text{cm}) = -1/(35\,\text{cm}) \Rightarrow f = -35\,\text{cm}$$

A negative power for a lens indicates a diverging lens. Now apply the same equation for an image distance of –15 cm and focal length of –35 cm to find the object distance.

$$s = \frac{s'f}{s' - f} = \frac{(-15\,\text{cm})(-35\,\text{cm})}{-15\,\text{cm} - (-35\,\text{cm})} = 26\,\text{cm}$$

So the closest obejct she can focus on with her glasses is 26 cm away.

Assess: She can't focus on things as close with her glasses as without them, so she may remove them for close work.

P19.53. Prepare: For telescopes the angular resolution is $\theta_1 = 1.22\lambda/D$.

We are given $\lambda = 550\,\text{nm}$ and $D = 2.4\,\text{m}$.

Solve:

$$\theta_1 = \frac{1.22\lambda}{D} = \frac{1.22(550 \times 10^{-9}\,\text{m})}{2.4\,\text{m}} = 2.8 \times 10^{-7}\,\text{rad}$$

Assess: The resolution of the Hubble Space Telescope is very good; we have received many fantastic photographs from it.

P19.57. Prepare: If two microtubules are next to each other, then their center-to-center separation is equal to their diameter: 25 nm. We will use Equation 19. to see if d_{min} is less than that.

We are given $NA = 1.4$ and $\lambda = 500\,\text{nm}$.

Solve:

$$d_{\text{min}} = \frac{0.61\lambda}{NA} = \frac{0.61(500\,\text{nm})}{1.4} = 218\,\text{nm}$$

This is about $10\times$ too big to see the microtubules. She would not be able to resolve two side-by-side.

Assess: These biological structures can be pretty tiny! Electron microscopes use electrons with shorter wavelength than light to see smaller objects and finer detail.

OPTICS

PptV.5. Reason: A nearsighted person has an eyeball that is too long, but can still focus on near objects. So where the distance from the cornea to the retina in the horse's eye is the longest is the answer. The answer is A.
Assess: Close objects focus farther away.

PptV.7. Reason: Close things that are at the bottom of the field of vision are in focus, so the feet will be in focus; but in the upper part of the field of vision distant objects are in focus, so the close head of the person will be out of focus. The answer is B.
Assess: See the previous two problems.

PptV.11. Reason: The rays reflected off the front surface undergo a 180° phase shift, which corresponds to half a wavelength. The rays reflected off the rear surface need to go another half wavelength farther to create constructive interference. Since the ray travels through the crystal layer twice the thickness needs to be $\lambda_n/4 = 330\ \text{nm}/4 = 80\ \text{nm}$.
The answer is A.
Assess: We need to use λ_n, not the wavelength in air.

PptV.13. Reason: Making sure the irises are wide open by taking the picture in dim light only enhances "eye shine" from the flash. The answer is A.
Assess: The other choices are reasonable strategies to minimize "eye shine."

PptV.17. Reason: The focal length is six times longer so the image will be six times taller, or 7.2 mm. The total light hitting the detector is the same, but is now spread out over 6^2 times the area, so the intensity is reduced by 36; the new intensity is $0.069\ \text{W/m}^2$.
Assess: As the text says, "if f is large, the image is large and its light is spread out and dim. If f is small, the image is small and the light is concentrated and bright."

20

ELECTRIC FIELDS AND FORCES

Q20.1. Reason: The statement of the question informs us that ball A is negative since charge is transferred to it from a plastic rod that has been rubbed with wool. Recall that a plastic rod rubbed with wool has a negative charge.

The first statement informs us that balls B, C, and D are either neutral or positive. If they are neutral, their attraction to ball A is due to charge polarization. If they are positive, their attraction to ball A is due to the fact that unlike charges attract.

The second statement informs us that balls B and D are neutral. If both are charged, there will be attraction or repulsion. If either one is charged, there will be attraction due to charge polarization of the other. Since there is neither attraction nor repulsion, neither is charged.

The last statement informs us that ball C must be charged. It is attracted to ball B due to polarization of charges on the neutral ball B.

Knowing that C is charged and attracted to A and that unlike charges attract, we can conclude that C has a positive charge.

To recap, ball A is negative, balls B and D are neutral, and ball C is positive.

Assess: By careful application of our knowledge of electrostatics we have been able to account for all observations. In some cases we had to combine two bits of information in order to make a conclusion regarding charge. All conclusions are consistent with the knowledge that like charges repel, unlike charges attract, and charge polarization can occur for a neutral object.

Q20.5. Reason: The negative charge is twice as far away from the dot as the positive charge is. But because Coulomb's law is an inverse square law the charge on the negative charge can't just be double, it must be $-4Q$.

Assess: At the dot the field due to the positive charge points toward the right and the field due to the negative charge points to the left; those vectors must sum to zero.

Q20.7. Reason: The following figure shows a representation of the charges on the metal sphere before and after the positively charged rod is brought near the neutral metal sphere.

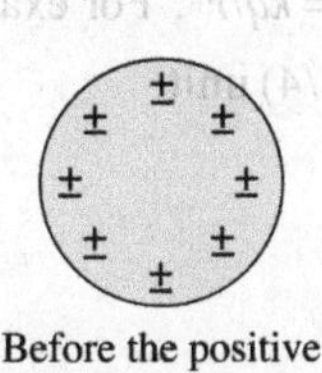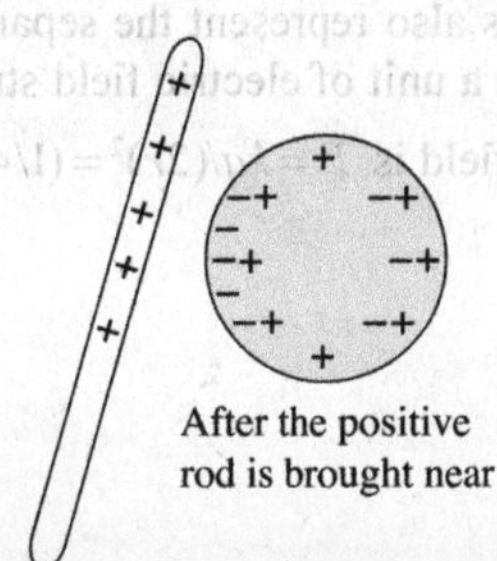

Before the positive
rod is brought near

After the positive
rod is brought near

(a) The metal sphere is a conductor, so the charges are free to move. Negative charges are attracted to the positive rod, but there are still equal numbers of positive and negative charges on the metal sphere since it is neutral.

In a real metal most of the inner electrons are still bound to their respective nucleus, but one or so of the outermost electrons from each atom become free to move around the metal. The nuclei don't move when the rod is broght near, but the free electrons do, leaving the far side of the sphere short of electrons and therefore positively charged.
(b) Since the negative charges are concentrated on the region of the sphere closest to the rod, there will be a net attractive force of the sphere to the rod.
Assess: This question uses our knowledge of charge polarization, the manner in which like and unlike charges interact, and the fact that the electrons are the mobile charge carriers.

Q20.9.   **Reason:**   Put the two metal spheres in contact with each other. Then rub the glass rod on the silk to charge the rod positively. Bring it near, but not touching, one of the spheres. This will make the near sphere negatively charged and the far sphere equally positively charged. Separate the spheres and remove the rod.
Assess: We don't want to touch the rod to the sphere and transfer charge, because it would be too easy to have both spheres end up positive.

Q20.13.   **Reason:**   Since like charges repel, negatively charged drug molecules should be placed near the negative (black) electrode to be pushed through the skin.
Assess: The field lines point from the red electrode (positive) to the black electrode (negative), but negative particles feel a force in the opposite direction from the local field lines.

Q20.17.   **Reason:**   Assume the distance between the particles is not changed. The force between two charged particles depends directly on the charge of each particle. If the charge on each is doubled, then the force is doubled twice. So the force increases by a factor of four.
Assess: This is true regardless of the sign of the two charges.

Q20.19.   **Reason:**   When lightning strikes, there is a tremendous transfer of charge from the cloud to the object struck.
When lightning strikes the metal plane, this charge is distributed over the surface of the plane.
There will initially be movement of charges over the plane (over a very short time interval) but a situation of static equilibrium will quickly be established.
Since we have established that there is no electric field inside a conductor, the passengers will experience no change.
Assess: In like manner, if you are inside a car struck by lighting, you will be safe. The tires will most likely be damaged as this tremendous amount of charge moves through them to ground.

Q20.21.   **Reason:**
(a) Yes, the field would be zero at a point on the line between the two charges, closer to the 10 nC charge.
(b) In this case the contributions from the two charges are in the same direction on the line between the charges, so there is no point between them at which the fields cancel each other.
Assess: In the first case the field contributions from the two charges are in opposite directions, so they can cancel out when the magnitudes are the same.

Q20.25.   **Reason:**   In the following figure, let's label the charges 1, 2, and 3 so we can keep track of the electric field due to each of them. Let's also represent the separation by r as shown in the figure. Next, in order to make a quantitative comparison, define a unit of electric field strength as $E = kq/r^2$. For example, at a distance of $2r$ from a charge q, the magnitude of the field is $E = kq/(2r)^2 = (1/4)(kq/r^2) = (1/4)$ unit.

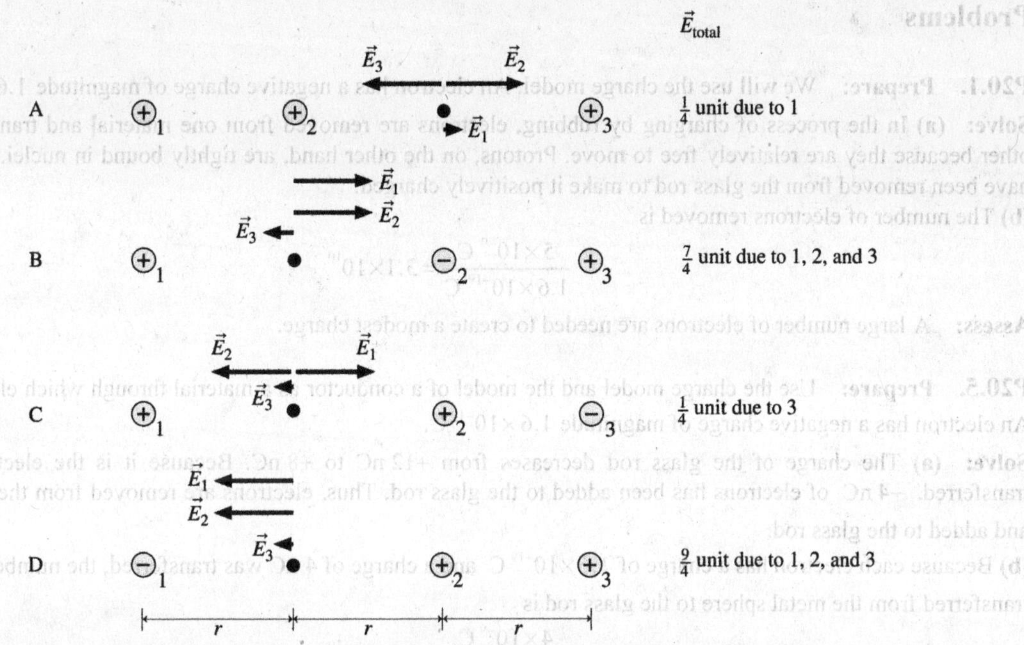

Note that in case A the electric field due to 2 and 3 add to zero and in case C the electric field due to 1 and 2 add to zero. As the figure points out, the electric field at the dot is greatest for case D.

Assess: In this case the calculation is simplified by defining a unit of electric field and then determining the electric field at the dot due to each charge in terms of this definition. This is a useful technique when you want to compare values.

Q20.27.   Reason: First, let's make sure we have a good mental picture of the physical situation. This is best done with a simple sketch as shown in the following figure.

Next determine the magnitude of the electric field due to each charge: $E_1 = kq_1/r_1^2 = 30 \times 10^4$ N/C to the right and $E_2 = kq_2/r_2^2 = 8 \times 10^4$ N/C to the right.

Since the electric field due to both charges is in the same direction, the magnitude of the resultant electric field is $E = 3.8 \times 10^5$ N/C and the direction of this field is to the right.

The correct answer is D.

Assess: The magnitude of this field is not out of line with other electric fields determined in the chapter.

Q20.29.   Reason: The positive end of the dipole is repelled by the lone positive charge, while the negative end of the dipole is attracted to the lone positive charge, so the dipole will initially rotate in a clockwise direction. The correct answer is A.

Assess: If the lone positive charge is fixed then the dipole would rotate clockwise 180 degrees and then rotate back 180 degrees counterclockwise.

Problems

P20.1.   Prepare: We will use the charge model. An electron has a negative charge of magnitude 1.6×10^{-19} C.

Solve: (a) In the process of charging by rubbing, electrons are removed from one material and transferred to the other because they are relatively free to move. Protons, on the other hand, are tightly bound in nuclei. So, electrons have been removed from the glass rod to make it positively charged.

(b) The number of electrons removed is

$$\frac{5\times10^{-9}\ \text{C}}{1.6\times10^{-19}\ \text{C}}=3.1\times10^{10}$$

Assess: A large number of electrons are needed to create a modest charge.

P20.5.   Prepare: Use the charge model and the model of a conductor as a material through which electrons move. An electron has a negative charge of magnitude 1.6×10^{-19} C.

Solve: (a) The charge of the glass rod decreases from +12 nC to +8 nC. Because it is the electrons that are transferred, -4 nC of electrons has been added to the glass rod. Thus, electrons are removed from the metal sphere and added to the glass rod.

(b) Because each electron has a charge of 1.6×10^{-19} C and a charge of 4 nC was transferred, the number of electrons transferred from the metal sphere to the glass rod is

$$\frac{4\times10^{-9}\ \text{C}}{1.6\times10^{-19}\ \text{C}}=2.5\times10^{10}$$

Assess: 25 billion electrons constitute a charge of 4 nC.

P20.7.   Prepare: Use the charge model and the model of a conductor as a material through which electrons move. An electron has a negative charge of magnitude 1.6×10^{-19} C.

Solve: Plastic is an insulator and does not transfer charge from one sphere to the other. The charge of metal sphere A is $(1.0\times10^{12})(-1.6\times10^{-19}\ \text{C})=-160$ nC and the charge of metal sphere B is 0 C.

Assess: Flow of charge does not occur between a charged conductor and an insulator when they are brought into contact.

P20.11.   Prepare: We need to solve Coulomb's law (Equation 20.1) for r:

$$F=K\frac{|q_1||q_2|}{r^2}$$

where $F=8.2\times10^{-4}$ N and $q_1=-5.0$ nC, $q_2=-12$ nC, and $K=9.0\times10^9\ \text{N}\cdot\text{m}^2/\text{C}^2$.

Solve:

$$r^2=K\frac{|q_1||q_2|}{F}$$

$$r=\sqrt{K\frac{|q_1||q_2|}{F}}=\sqrt{(9.0\times10^9\ \text{N}\cdot\text{m}^2/\text{C}^2)\frac{|-0.5\times10^{-9}\ \text{C}||-12\times10^{-9}\ \text{C}|}{8.2\times10^{-4}\ \text{N}}}=0.026\ \text{m}=2.6\ \text{cm}$$

Assess: Notice the N and C cancel out leaving units of m.

P20.15.   Prepare: The charged particles are point charges. The charge q_2 is in static equilibrium, so the net force on q_2 is zero. If q_2 is positive, q_1 will have to be positive to make the net force zero on q_2. And, if q_2. is negative, q_1 will still have to be positive for q_2 to be in equilibrium. We will assume that the charge q_2 is positive. For this situation, the force on q_2 by the -2 nC charge is to the left and by q_1 is to the right.

Solve: We have

$$\vec{F}_{\text{net on }q_2}=\vec{F}_{q_1\text{ on }q_2}+\vec{F}_{-2\text{ nC on }q_2}=\left(\frac{1}{4\pi\varepsilon_0}\frac{|q_1||q_2|}{(0.2\ \text{m})^2},+x\text{-direction}\right)+\left(\frac{1}{4\pi\varepsilon_0}\frac{(2\times10^{-9}\ \text{C})|q_2|}{(0.10\ \text{m})^2},-x\text{-direction}\right)=0\ \text{N/C}$$

Thus,

$$\frac{q_1}{(0.2\ \text{m})^2} - \frac{2\times10^{-9}\ \text{C}}{(0.10\ \text{m})^2} = 0\ \text{N/C} \Rightarrow q_1 = 8.0\ \text{nC}$$

Assess: If the charge q_2 is assumed negative, the force on q_2 by the $-2\ \text{nC}$ charge is to the right and by q_1 is to the left. The magnitude of q_1 remains unchanged.

P20.17. Prepare: Assume the glass bead, the proton, and the electron are point charges. A visual overview of the forces and the coordinate system is shown.

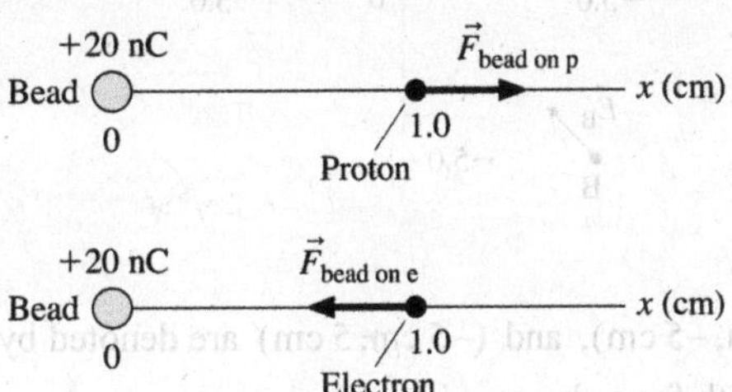

Solve: Coulomb's law gives

$$F_{\text{bead on electron}} = F_{\text{bead on proton}} = \frac{(9\times10^{9}\ \text{N}\cdot\text{m}^2/\text{C}^2)(20\times10^{-9}\ \text{C})(1.60\times10^{-19}\ \text{C})}{(1.0\times10^{-2}\ \text{m})^2} = 2.88\times10^{-13}\ \text{N}$$

(a) Newton's second law is $F = ma$, so

$$a_{\text{proton}} = \frac{F_{\text{bead on proton}}}{m_{\text{proton}}} = \frac{2.88\times10^{-13}\ \text{N}}{1.67\times10^{-27}\ \text{kg}} = 1.7\times10^{14}\ \text{m/s}^2$$

In vector form

$$\vec{a}_{\text{proton}} = (1.7\times10^{14}\ \text{m/s}^2,\ \text{away from bead})$$

(b) Similarly,

$$a_{\text{electron}} = \frac{F_{\text{bead on electron}}}{m_{\text{electron}}} = \frac{2.88\times10^{-13}\ \text{N}}{9.11\times10^{-31}\ \text{kg}} = 3.2\times10^{17}\ \text{m/s}^2$$

Thus $\vec{a}_{\text{electron}} = (3.2\times10^{17}\ \text{m/s}^2,\ \text{toward bead})$.

Assess: Because of their small mass, electrons and protons are accelerated tremendously by electric fields.

P20.21. Prepare: Protons and electrons are point charges and produce electric fields, according to Equation 20.6. The charge on a proton is positive and on an electron is negative.

Solve: **(a)** The electric field of the proton is

$$\vec{E} = \left(\frac{1}{4\pi\varepsilon_0}\frac{|q|}{r^2},\ \text{away from } q\right) = \left((9.0\times10^{9}\ \text{N}\cdot\text{m}^2/\text{C}^2)\left[\frac{+1.6\times10^{-19}\ \text{C}}{(1.0\times10^{-3}\ \text{m})^2}\right],\ \text{away from } q\right)$$

$$= (1.4\times10^{-3}\ \text{N/C},\ \text{away from proton})$$

(b) The electric field of the electron is

$$\vec{E} = \left(\frac{1}{4\pi\varepsilon_0}\frac{|q|}{r^2},\ \text{toward } q\right) = \left((9.0\times10^{9}\ \text{N}\cdot\text{m}^2/\text{C}^2)\left[\frac{+1.6\times10^{-19}\ \text{C}}{(1.0\times10^{-3}\ \text{m})^2}\right],\ \text{toward } q\right)$$

$$= (1.4\times10^{-3}\ \text{N/C},\ \text{toward electron})$$

Assess: An electron charge is very small, so its electric field at a point 1 mm away was expected to be small.

P20.23.   Model: The electric field is that of a negative charge located at the origin as shown below. We will use Equation 20.6.

The positions $(0\text{ cm}, 5\text{ cm})$, $(-5\text{ cm}, -5\text{ cm})$, and $(-5\text{ cm}, 5\text{ cm})$ are denoted by A, B, and C, respectively.

Solve: (a) The electric field strength for a charge q is

$$E = K\frac{|q|}{r^2}$$

Using $K = 9.0\times10^9\,\text{N}\cdot\text{m}^2/\text{C}^2$ and $|q| = 10.0\times10^{-9}\,\text{C}$,

$$E = \frac{90.0\,\text{N}\cdot\text{m}^2/\text{C}}{r^2}$$

The electric field strengths at points A, B, and C are

$$E_A = \frac{90.0\,\text{N}\cdot\text{m}^2/\text{C}}{(5.0\times10^{-2}\,\text{m})^2} = 3.6\times10^4\,\text{N/C}$$

$$E_B = \frac{90.0\,\text{N}\cdot\text{m}^2/\text{C}}{(-5.0\times10^{-2}\,\text{m})^2 + (5.0\times10^{-2}\,\text{m})^2} = 1.8\times10^4\,\text{N/C}$$

$$E_C = \frac{90.0\,\text{N}\cdot\text{m}^2/\text{C}}{(-5.0\times10^{-2}\,\text{m})^2 + (-5.0\times10^{-2}\,\text{m})^2} = 1.8\times10^4\,\text{N/C}$$

(b) The three vectors are shown in the diagram.

Assess: Note that the vectors $\vec{E}_A$, $\vec{E}_B$, and $\vec{E}_C$ are pointing toward the negative charge.

P20.25.   Prepare: The electric field is that of the two charges placed on the y-axis. We denote the upper charge by q_1 and the lower charge by q_2. The electric field at the dot due to the positive charge is directed away from the charge and making an angle of $45°$ below the $+x$ axis, but the electric field due to the negative charge is directed toward it making an angle of $45°$ below the $-x$ axis.

Solve: The electric field strength of q_1 is

$$E_1 = K\frac{|q_1|}{r_1^2} = \frac{(9.0\times10^9\,\text{N}\cdot\text{m}^2/\text{C}^2)(1\times10^{-9}\,\text{C})}{(0.050\,\text{m})^2 + (0.050\,\text{m})^2} = 1800\,\text{N/C}$$

Similarly, the electric field strength of q_2 is

$$E_2 = K\frac{|q_2|}{r_2^2} = \frac{(9.0\times10^9\,\text{N}\cdot\text{m}^2/\text{C}^2)(1\times10^{-9}\,\text{C})}{(0.050\,\text{m})^2 + (0.050\,\text{m})^2} = 1800\,\text{N/C}$$

We will now calculate the components of these electric fields. The electric field due to q_1 is away from q_1 in the fourth quadrant and that due to q_2 is toward q_2 in the third quadrant. Their components are

$$E_{1x} = E_1 \cos 45°$$
$$E_{1y} = -E_1 \sin 45°$$
$$E_{2x} = -E_2 \cos 45°$$
$$E_{2y} = -E_2 \sin 45°$$

The x and y components of the net electric field are:

$$(E_{net})_x = E_{1x} + E_{2x} = E_1 \cos 45° - E_2 \cos 45° = 0 \text{ N/C}$$
$$(E_{net})_y = E_{1y} + E_{2y} = -E_1 \sin 45° - E_2 \sin 45° = -2500 \text{ N/C}$$

$$\Rightarrow \vec{E}_{net \, at \, dot} = (2500 \text{ N/C, along } -y \text{ axis})$$

Thus, the strength of the electric field is 2500 N/C and its direction is vertically downward.

Assess: A quick visualization of the components of the two electric fields shows that the horizontal components cancel.

P20.29. Prepare: The electric field is uniform in a region of space between closely spaced capacitor plates and is given by Equation 20.7.

Solve: The electric field inside a capacitor is $E = Q/(\varepsilon_0 A)$. Thus, the charge needed to produce a field of strength E is

$$Q = \varepsilon_0 A E = (8.85 \times 10^{-12} \text{ C}^2/\text{N} \cdot \text{m}^2)(0.04 \text{ m} \times 0.04 \text{ m})(1.0 \times 10^6 \text{ N/C}) = 14 \text{ nC}$$

Thus, one plate has a charge of 14 nC and the other has a charge of -14 nC.

Assess: Note that the capacitor as a whole has no net charge.

P20.31. Prepare: For the negatively charged bead to be suspended at rest the net force on it must be zero. There is a downward gravitational force, so there must be an upward electric force.

Solve: (a) For the negatively charged bead to be suspended it must be repelled by a negative plate below and attracted to a positive plate above. So the upper plate is positively charged.

(b) The electric force has the same magnitude as the gravitational force: $qE = mg$, where q is the 6.0 nC charge on the bead and $E = Q/(\varepsilon_0 A)$. We seek Q, the charge on the upper plate.

$$qE = mg \Rightarrow E = \frac{Q}{\varepsilon_0 A} = \frac{mg}{q} \Rightarrow$$

$$Q = \frac{(\varepsilon_0 A)(mg)}{q} = \frac{(8.85 \times 10^{-12} \text{ C}^2/\text{N} \cdot \text{m}^2) \pi (6.0 \text{ cm})^2 (1.0 \text{ g})(9.8 \text{ m/s}^2)}{6.0 \text{ nC}} = 1.6 \times 10^{-7} \text{ C}$$

Assess: This is a reasonable charge for a parallel plate capacitor.

P20.35. Prepare: Equation 20.8 tells us the force on a charged object in an electric field: $\vec{F}_{on \, q} = q\vec{E}$. We are given $q = 30e$ and $E = 1500$ N/C.

Solve:

$$F_{on \, q} = qE = (30)(1.6 \times 10^{-19} \text{ C})(1500 \text{ N/C}) = 7.2 \times 10^{-15} \text{ N}$$

Assess: Notice the C's cancel out leaving units of N. The answer is very small, but that is what we expect for such a small charge.

P20.39. Prepare: Use the charge model. The number of electrons per atom is the atomic number, and both the atomic number (29) and the average atomic mass (63.5 g) are taken from the periodic table in the textbook.

Solve: The number of moles in the penny is

$$n = \frac{M}{A} = \frac{3.1 \text{ g}}{63.5 \text{ g/mol}} = 0.04882 \text{ mol}$$

The number of copper atoms in the penny is

$$N = nN_A = (0.04882 \text{ mol})(6.02 \times 10^{23} \text{ mol}^{-1}) = 2.939 \times 10^{22}$$

Since each copper atom has 29 electrons and 29 protons, the total positive charge in the copper penny is

$$(29 \times 2.939 \times 10^{22})(1.60 \times 10^{-19} \text{ C}) = 1.4 \times 10^5 \text{ C}$$

Similarly, the total negative charge is $-1.4 \times 10^5 \, \text{C}$.
Assess: Total positive and negative charges are equal in magnitude.

P20.41. **Prepare:** The ^{125}Xe nucleus and the proton are point charges. That is, all the charge on the Xe nucleus is assumed to be at its center. Electron charge is $1.60 \times 10^{-16} \, \text{C}$.

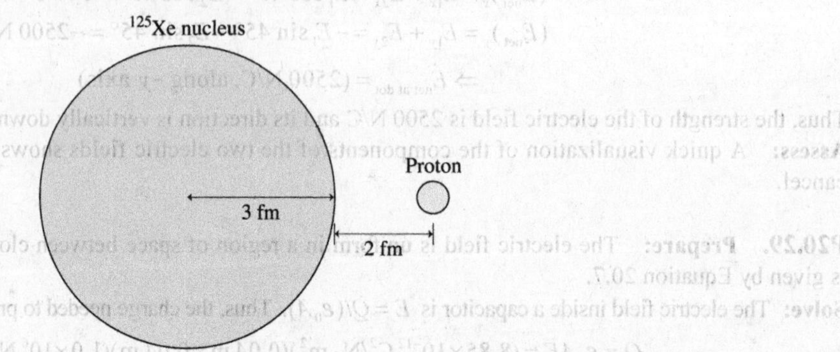

Solve: **(a)** The magnitude of the force between the nucleus and the proton is given by Coulomb's law:

$$F_{\text{nucleus on proton}} = \frac{K|q_{\text{nucleus}}||q_{\text{proton}}|}{r^2} = \frac{(9.0 \times 10^9 \, \text{N} \cdot \text{m}^2/\text{C}^2)(54 \times 1.60 \times 10^{-19} \, \text{C})(1.60 \times 10^{-19} \, \text{C})}{(5.0 \times 10^{-15} \, \text{m})^2} = 498 \, \text{N}$$

which we will report as 500 N.
(b) Applying Newton's second law to the proton,

$$F_{\text{on proton}} = m_{\text{proton}} a_{\text{proton}} \Rightarrow a_{\text{proton}} = \frac{498 \, \text{N}}{1.67 \times 10^{-27} \, \text{kg}} = 2.98 \times 10^{29} \, \text{m/s}^2$$

which we will report as $3.0 \times 10^{29} \, \text{m/s}^2$.
Assess: A relatively large force on such a small object as a proton will cause tremendous acceleration.

P20.43. **Prepare:** Positively charged objects A and B are point charges. The repulsive force by A on B and by B on A are equal and given by Coulomb's law.

$\vec{F}_{\text{B on A}}$ ← $r = 10$ cm → $\vec{F}_{\text{A on B}}$

(+) A   (+) B

$m_A = 0.10 \, \text{kg}$ $m_B = 0.10 \, \text{kg}$

$q_A = 2q_B$ q_B

Solve: **(a)** It is given that $F_{\text{A on B}} = 0.45 \, \text{N}$. By Newton's third law, $F_{\text{B on A}} = F_{\text{A on B}} = 0.45 \, \text{N}$.

(b) Coulomb's law is

$$F_{\text{B on A}} = F_{\text{A on B}} = 0.45 \, \text{N} = \frac{Kq_A q_B}{r^2} = \frac{K(2q_B)(q_B)}{r^2}$$

$$\Rightarrow q_B = \sqrt{\frac{(0.45 \, \text{N})r^2}{2K}} = \sqrt{\frac{(0.45 \, \text{N})(10 \times 10^{-2} \, \text{m})^2}{2(9.0 \times 10^9 \, \text{N} \cdot \text{m}^2/\text{C}^2)}} = 5.0 \times 10^{-7} \, \text{C} \Rightarrow q_A = 2q_B = 1.0 \times 10^{-6} \, \text{C}$$

Assess: A relatively large force of 0.45 N between charges separated by 10 cm must mean significant charge on the objects.

P20.45. Prepare: The electric field is that of the two 1 nC charges located on the y-axis. We denote the top 1 nC charge by q_1 and the bottom 1 nC charge by q_2. The electric fields ($\vec{E}_1$ and $\vec{E}_2$) of both the positive charges are directed away from their respective charges. With vector addition, they yield the net electric field $\vec{E}_{net}$ at the point P indicated by the dot.

Solve: The electric fields from q_1 and q_2 are

$$\vec{E}_1 = \left(K\frac{|q_1|}{r_1^2}, \text{along} + x\text{-axis} \right) = \left(\frac{(9.0 \times 10^9 \text{ N.m}^2/\text{C}^2)(1 \times 10^{-9} \text{ C})}{(0.05 \text{ m})^2}, \text{along} + x\text{-axis} \right)$$

$$= (3600 \text{ N/C, along} + x\text{-axis})$$

$$\vec{E}_2 = \left(\frac{1}{4\pi\varepsilon_0} \frac{|q_2|}{r_2^2}, \theta \text{ above} + x\text{-axis} \right) = (720 \text{ N/C}, \theta \text{ above} + x\text{-axis})$$

Because $\tan\theta = 10 \text{ cm}/5 \text{ cm}$, $\theta = \tan^{-1}(2) = 63.43°$.

We will now calculate the components of these electric fields. The electric field due to q_1 is away from q_1 along $+x$ and that due to q_2 is away from q_2 in the first quadrant. Their components are

$$E_{1x} = E_1$$
$$E_{1y} = 0$$
$$E_{2x} = E_2 \cos 63.45°$$
$$E_{2y} = E_2 \sin 63.45°$$

The x and y components of the net electric field are:

$$(E_{net})_x = E_{1x} + E_{2x} = E_1 + E_2 \cos 63.45° = 3922 \text{ N/C}$$
$$(E_{net})_y = E_{1y} + E_{2y} = 0 + E_2 \sin 63.45° = 644 \text{ N/C}$$

Thus, the strength of the electric field at P is

$$E_{net} = \sqrt{(3922 \text{ N/C})^2 + (644 \text{ N/C})^2} = 3975 \text{ N/C}$$

which will be reported as 4000 N/C.

To find the angle this net vector makes with the x-axis, we calculate

$$\tan\phi = \frac{644 \text{ N/C}}{3922 \text{ N/C}} \Rightarrow \phi = 9.3°$$

Assess: Because of the inverse square dependence on distance, $E_2 < E_1$. Additionally, because the point P has no special symmetry relative to the charges, we expected the net field to be at an angle relative to the x-axis.

P20.51. Prepare: The charges are point charges. Placing the 1 nC charge at the origin and calling it q_1, the -6 nC is q_3, the q_2 charge is in the first quadrant, and the q_4 charge is in the second quadrant. The net electric force on q_1 is the vector sum of the electric forces from the other three charges q_2, q_3, and q_4.

Solve: We have

$$\vec{F}_{2 \text{ on } 1} = \left(\frac{K|q_1||q_2|}{r^2}, \text{ away from } q_2 \right)$$

$$= \left(\frac{(9.0 \times 10^9 \text{ N} \cdot \text{m}^2/\text{C}^2)(1 \times 10^{-9} \text{ C})(2 \times 10^{-9} \text{ C})}{(5.0 \times 10^{-2} \text{ m})^2}, \text{ away from } q_2 \right)$$

$$= (0.72 \times 10^{-5} \text{ N, away from } q_2)$$

$$\vec{F}_{3\,\text{on}\,1} = \left(\frac{K|q_1||q_3|}{r^2},\ \text{toward}\ q_3\right)$$

$$= \left(\frac{(9.0\times10^9\ \text{N}\cdot\text{m}^2/\text{C}^2)(1\times10^{-9}\ \text{C})(6\times10^{-9}\ \text{C})}{(5.0\times10^{-2}\ \text{m})^2},\ \text{toward}\ q_3\right)$$

$$= (2.16\times10^{-5}\ \text{N},\ \text{away from}\ q_3)$$

$$\vec{F}_{4\,\text{on}\,1} = \left(\frac{K|q_1||q_4|}{r^2},\ \text{away from}\ q_4\right) = (0.72\times10^{-5}\ \text{N},\ \text{away from}\ q_4)$$

$$(\vec{F}_{2\,\text{on}\,1})_x = -(0.72\times10^{-5}\ \text{N})(\cos 45°) = -(0.509\times10^{-5}\ \text{N})$$

$$(\vec{F}_{2\,\text{on}\,1})_y = -(0.72\times10^{-5}\ \text{N})(\sin 45°) = -(0.509\times10^{-5}\ \text{N})$$

$$(\vec{F}_{3\,\text{on}\,1})_x = 0\ \text{N}$$

$$(\vec{F}_{3\,\text{on}\,1})_y = (2.16\times10^{-5}\ \text{N})$$

$$(\vec{F}_{4\,\text{on}\,1})_x = (0.72\times10^{-5}\ \text{N})(\cos 45°) = (0.509\times10^{-5}\ \text{N})$$

$$(\vec{F}_{4\,\text{on}\,1})_y = -(0.72\times10^{-5}\ \text{N})(\sin 45°) = -(0.509\times10^{-5}\ \text{N})$$

$$(\vec{F}_{\text{on}\,1})_x = (\vec{F}_{2\,\text{on}\,1})_x + (\vec{F}_{3\,\text{on}\,1})_x + (\vec{F}_{4\,\text{on}\,1})_x = 0\ \text{N}$$

$$(\vec{F}_{\text{on}\,1})_y = (\vec{F}_{2\,\text{on}\,1})_y + (\vec{F}_{3\,\text{on}\,1})_y + (\vec{F}_{4\,\text{on}\,1})_y = 1.14\times10^{-5}\ \text{N}$$

So the force on the 1 nC charge is $1.1\times10^{-5}\ \text{N}$ directed vertically up.

P20.53. Prepare: The charged particles are point charges. The two 2 nC charges exert an upward force on the 1 nC charge. Since the net force on the 1 nC charge is zero, the unknown charge must exert a downward force of equal magnitude. This implies that q is a positive charge.

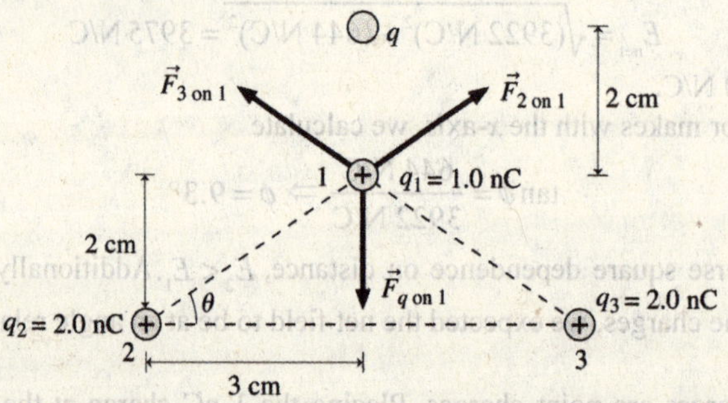

Solve: The force of charges 2 on charge 1 is

$$\vec{F}_{2\,\text{on}\,1} = \left(\frac{K|q_1||q_2|}{r_{12}^2},\ \text{away from}\ q_2\right)$$

$$= \left(\frac{(9.0\times10^9\ \text{N}\cdot\text{m}^2/\text{C}^2)(1.0\times10^{-9}\ \text{C})(2.0\times10^{-9}\ \text{C})}{(0.020\ \text{m})^2 + (0.030\ \text{m})^2},\ \text{away from}\ q_2\right)$$

$$= (1.385\times10^{-5}\ \text{N},\ \text{away from}\ q_2)$$

From the figure, $\theta = \tan^{-1}(2/3) = 33.69°$. So,

$$(\vec{F}_{2\,\text{on}\,1})_x = (1.385\times10^{-5}\ \text{N})(\cos 33.69°) = (1.152\times10^{-5}\ \text{N})$$

$$(\vec{F}_{2\,\text{on}\,1})_y = (1.385\times10^{-5}\ \text{N})(\sin 33.69°) = (0.768\times10^{-5}\ \text{N})$$

From symmetry, $\vec{F}_{3\,\text{on}\,1}$ is the same except the x-component is reversed. So, when we add $\vec{F}_{2\,\text{on}\,1}$ and $\vec{F}_{3\,\text{on}\,1}$, the x-components cancel and the y-components add to give

$$\vec{F}_{2\,\text{on}\,1}+\vec{F}_{3\,\text{on}\,1}=1.536\times10^{-5}\,\text{N}$$

$\vec{F}_{q\,\text{on}\,1}$ must have the same magnitude, pointing in the vertically downward direction, so

$$\vec{F}_{q\,\text{on}\,1}=1.536\times10^{-5}\,\text{N}=\frac{K|q||q_1|}{r^2}\Rightarrow q=\frac{(1.536\times10^{-5}\,\text{N})(0.020\,\text{m})^2}{(9.0\times10^9\,\text{N}\cdot\text{m}^2/\text{C}^2)(1.0\times10^{-9}\,\text{C})}=0.68\,\text{nC}$$

A positive charge $q=0.68\,\text{nC}$ will cause the net force on the 1 nC charge to be zero.

P20.55. Prepare: The charges are point charges. We will denote the charges counterclockwise starting at the upper left by 1, 2, and 3, respectively.

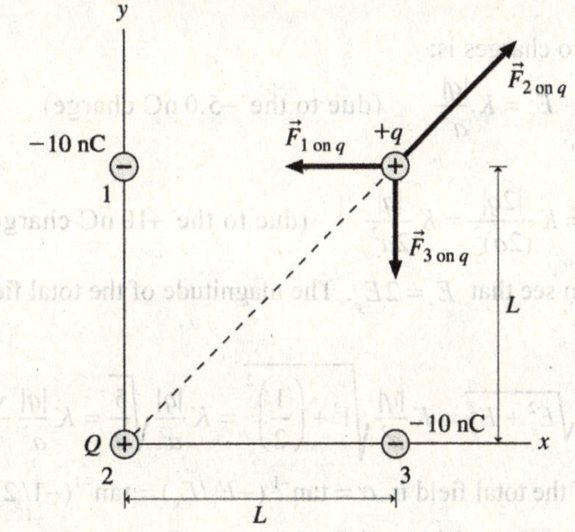

There is enough symmetry in this problem that we can simply the calculation. We need to show that in the x-direction $F_{1\,\text{on}\,q}=-F_{2\,\text{on}\,q}\cos 45°$ and solve for Q. We do not need to re-do the calculation in the y-direction as the result would be the same because of the symmetry.

Solve: First write the force of charge 1 on q and then find the x-component of it

$$\vec{F}_{1\,\text{on}\,q}=\left(\frac{K|-10\,\text{nC}||q|}{L^2},\ \text{toward}-Q\right)$$

$$(\vec{F}_{1\,\text{on}\,q})_x=-\frac{KQq}{L^2}$$

Now do the same for the force of charge 2 (i.e., Q) on q. The distance is now $\sqrt{2}L$.

$$\vec{F}_{2\,\text{on}\,q}=\left(\frac{K|Q||q|}{(\sqrt{2}L)^2},\ \text{away from}\ 4Q\right)$$

$$(\vec{F}_{2\,\text{on}\,q})_x=\frac{KQq}{2L^2}\cos 45°$$

Set the sum of the x-components to zero.

$$(\vec{F}_{1\,\text{on}\,q})_x=-(\vec{F}_{2\,\text{on}\,q})_x$$

$$-\frac{K(-10\,\text{nC})q}{L^2}=-\frac{KQq}{2L^2}\cos 45°$$

Cancel the common factors and solve for Q.

$$(10\,\text{nC})=\frac{Q}{2}(\cos 45°)=\frac{Q}{2}\left(\frac{\sqrt{2}}{2}\right)\Rightarrow Q=\frac{4}{\sqrt{2}}(10\,\text{nC})=28\,\text{nC}$$

This same Q will also do the job in the y-direction to cancel the force from charge 3.

Assess: Because the x- and y-components of the net force are equal, the net force is away from $+Q$ connecting it to $+q$.

P20.59. Prepare: The field due to the third charge $(-10\,\text{nC})$ must be equal in magnitude and opposite in direction to the vector sum of the contributions to the field at the origin by the other two charges. We'll use Equation 20.6 for the field from a point charge:

$$E = \frac{K|q|}{r^2}$$

with the direction away from q if $q > 0$ and toward q if $q < 0$.

This problem is most elegantly done if we use letters to represent the values; this will save a lot of writing, and will actually make the algebra easier. Say that $q = 5.0\,\text{nC}$ so that the charge on the x-axis is $-q$ and the charge on the y-axis is $2q$. Also let a stand for the distance to the closest charge, 5.0 cm; this means the distance to the $+10\,\text{nC}$ charge is $2a$.

Solve: The field due to the first two charges is:

$$E_x = K\frac{|q|}{a^2} \qquad \text{(due to the } -5.0\,\text{nC charge)}$$

in the positive x-direction; and

$$E_y = K\frac{|2q|}{(2a)^2} = K\frac{|q|}{2a^2} \qquad \text{(due to the } +10\,\text{nC charge)}$$

in the negative y-direction. You can see that $E_x = 2E_y$. The magnitude of the total field due to the first two charges is then

$$E = \sqrt{E_x^2 + E_y^2} = K\frac{|q|}{a^2}\sqrt{1^2 + \left(\frac{1}{2}\right)^2} = K\frac{|q|}{a^2}\sqrt{\frac{5}{4}} = K\frac{|q|}{a^2}\frac{\sqrt{5}}{2}$$

Because $E_x = 2E_y$, the direction of the total field is $\alpha = \tan^{-1}(-E_y/E_x) = \tan^{-1}(-1/2) = -26.57°$.

We now have the magnitude and direction of the total field from the first two charges.

The field due to the third charge must have the same magnitude and opposite direction. The third charge has charge $-2q$, and is r_3 away from the origin. We need to solve for r_3:

$$E_3 = K\frac{|q|}{a^2}\frac{\sqrt{5}}{2} = K\frac{|-2q|}{r_3^2}$$

This implies that

$$\frac{1}{a^2}\frac{\sqrt{5}}{2} = \frac{2}{r_3^2}$$

$$r_3^2 = \frac{4a^2}{\sqrt{5}}$$

$$r_3 = \frac{2a}{\sqrt[4]{5}}$$

The direction of E_3 is $\theta = \alpha + 180° = 153.43°$.

We now have the distance from the origin and direction of the third charge; we can use these to get the x- and y-coordinates.

$$x = r_3\cos\theta = \frac{2q}{\sqrt[4]{5}}\cos\theta = \frac{2(5.0\text{ cm})}{\sqrt[4]{5}}\cos 153.43° = -5.98\text{ cm} \approx -6.0\text{ cm}$$

$$y = r_3\sin\theta = \frac{2a}{\sqrt[4]{5}}\sin\theta = \frac{2(5.0\text{ cm})}{\sqrt[4]{5}}\sin 153.43° = 2.00\text{ cm} \approx 3.0\text{ cm}$$

The final result is that (x, y) for the $-10\,\text{nC}$ charge is $(-6.0\text{ cm}, 3.0\text{ cm})$.

Assess: A quick sketch of the coordinate plane with the three charges and their field contributions at the origin should convince you that we have the right answer.

The fact that we never need to plug in the actual value of q anywhere shows that result is independent of q as long as the ratios of the q's are the same. Furthermore, since K also canceled, then the result would hold for any inverse-square field, such as a gravitational field if you could find both positive and negative masses.

P20.63. Prepare: Just as the bead is lifted off the table the normal force from the table will be zero. The free-body diagram will have only the downward gravitational force and the upward (because the charges have opposite sign) electric force, and they will be equal in magnitude.

Solve: Set the magnitude of the gravitational force equal to the magnitude of the electric force and solve for r.

$$k\frac{|q_1||q_2|}{r^2}=mg \Rightarrow r=\sqrt{\frac{k|q_1||q_2|}{mg}}=\sqrt{\frac{(8.99\times10^9\ \text{N}\cdot\text{m}^2/\text{C}^2)(2.5\ \text{nC})(5.6\ \text{nC})}{(4.0\ \text{mg})(9.8\ \text{m/s}^2)}}=5.7\ \text{cm}$$

Assess: This is a reasonable distance between beads.

P20.65. Prepare: The charged ball attached to the string is a point charge. The ball is in static equilibrium in the external electric field when the string makes an angle $\theta=20°$ with the vertical. The three forces acting on the charged ball are the electric force due to the field, the weight of the ball, and the tension force.

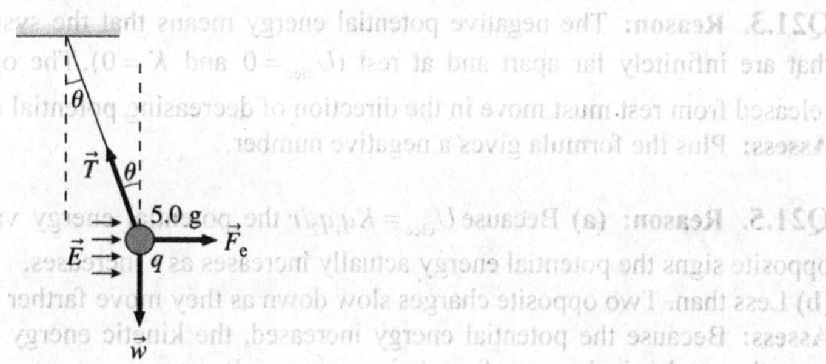

Solve: In static equilibrium, Newton's second law for the ball is $\vec{F}_{\text{net}}=\vec{T}+\vec{w}+\vec{F}_e=\vec{0}$. In component form,

$$(F_{\text{net}})_x=T_x+0\ \text{N}+qE=0\ \text{N} \qquad (F_{\text{net}})_y=T_y-mg+0\ \text{N}=0\ \text{N}$$

The two previous equations simplify to

$$T\sin\theta=qE \qquad T\cos\theta=mg$$

Divide the first equation by the second to get

$$\tan\theta=\frac{qE}{mg} \Rightarrow q=\frac{mg\tan\theta}{E}=\frac{(5.0\times10^{-3}\ \text{kg})(9.8\ \text{N/kg})\tan 20°}{100\,000\ \text{n/C}}=1.78\times10^{-7}\ \text{C}=180\ \text{nC}$$

to two significant figures.

P20.67. Prepare: The acceleration due to gravity is of comparable size to the acceleration given in the problem, so we will not neglect gravity. The forces on the bead are mg down and qE up. Use Newton's second law.

Solve:

$$\sum F=qE-mg=ma \Rightarrow q=\frac{m(a+g)}{E}=\frac{(0.0010\ \text{kg})(20\ \text{m/s}^2+9.8\ \text{m/s}^2)}{200\,000\ \text{N/C}}=150\ \text{nC}$$

Assess: The answer is in the ballpark of charges for plastic beads.

P20.69. Prepare: When the bead is at rest the sum of the forces on it is zero, so the magnitude of the force from the $4q$ charge at the right must be the same as the magnitude of the force from the q charge at the left. This can be solved with Coulomb's law and some algebra, but it is easier to just reason about it.

Solve: Since Coulomb's law is an inverse square law, the $4q$ charge needs to be twice as far away from the free-to-move charge as the q on the left is. The place that is twice as far away from $x=4.0$ cm as the origin is one-third the way from the origin, or at $x=1.3$ cm.

Assess: The charge that is twice as far away has four times the charge to produce the same magnitude force.

ELECTRIC POTENTIAL

Q21.3. Reason: The negative potential energy means that the system has less energy than two charged particles that are infinitely far apart and at rest ($U_{elec} = 0$ and $K = 0$). The opposite charges are a bound system. A charge released from rest must move in the direction of decreasing potential energy.
Assess: Plus the formula gives a negative number.

Q21.5. Reason: (a) Because $U_{elec} = Kq_1q_2/r$ the potential energy varies as $1/r$, but because the two charges have opposite signs the potential energy actually increases as r increases.
(b) Less than. Two opposite charges slow down as they move farther apart.
Assess: Because the potential energy increased, the kinetic energy decreased (we assume there was no friction to keep the mechanical energy from being conserved).

Q21.7. Reason: (a) The electric field between the plates is uniform and since both protons move the same distance in a direction opposite that of the electric field, they both experience the same electric potential difference ($\Delta V = E\Delta d$) and hence the same change in electric potential energy ($\Delta U = q\Delta V = qE\Delta d$).
(b) At point 1, the protons have the same electric potential energy and the same kinetic energy (since they are at the same position and have the same mass and speed). Since they both experience the same increase in electric potential energy, they will both experience the same decrease in kinetic energy. This means that the kinetic energy of the proton at point 2 will be equal to the kinetic energy of the proton at point 3, hence that have the same speed.
Assess: This question does a nice job of showing that potential energy and kinetic energy are tied together in a conservative force system.

Q21.11. Reason: The information critical to the solution of this question is the relationship between electric potential, the source charge causing this potential, and the distance from this source charge ($V = kQ/r$). Since (for this question) the electric potential is due to a negative charge, all electric potentials are negative. Since positions 2, 3, and 4 are equidistant from the negative charge, they all have the same electric potential. Since positions 1 and 5 are equidistant from the negative charge, they both have the same electric potential. Since positions 1 and 5 are further from the charge than positions 2, 3, and 4, they will have a smaller negative electric potential. Ranking the electric potentials from most positive to most negative, obtain

$$V_1 = V_5 > V_2 = V_3 = V_4$$

most positive   least positive
least negative   most negative

Assess: To solve this problem, we have just carefully listed the information contained in the expression for the electric potential.

Q21.15. Reason: Since the figure gives information about the equipotential lines and shows the distance between these lines, we can use $E = \Delta V/\Delta d$ to get a qualitative value for E at each site. A qualitative value is adequate for comparison purposes. In order for us to talk about the different potential differences (ΔV) and distances (Δd) let's label them as shown in the figure.

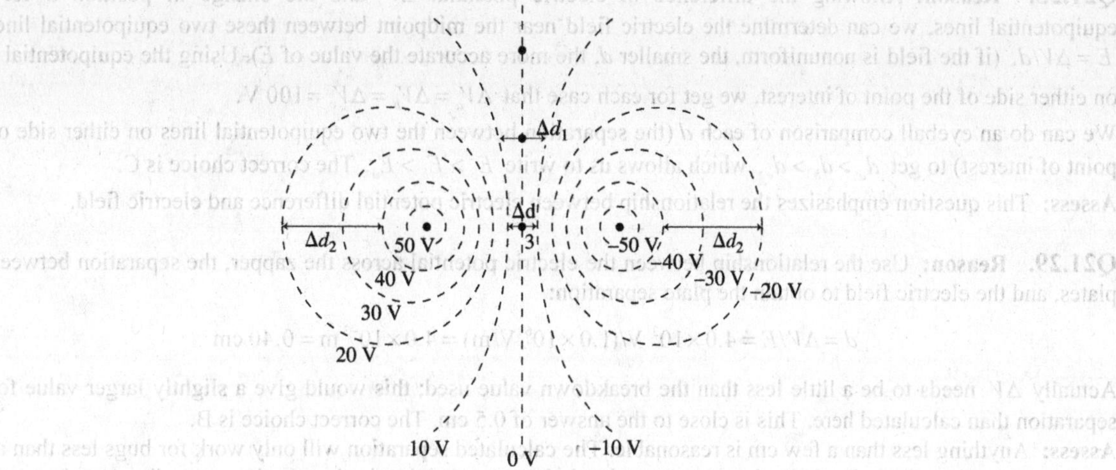

Notice that each ΔV (between the first equipotential line on either side) is 20 volts: $\Delta V_1 = \Delta V_2 = \Delta V_3 = \Delta V_4 = 20$ V. Looking at the Δd for each case we see that

$$\Delta d_2 = \Delta d_4 > \Delta d_1 > \Delta d_3$$

Using $E = \Delta V/\Delta d$, this allows us to conclude that

$$E_3 < E_1 < E_2 = E_4$$

Assess: Since the electric field is not uniform, the expression $E = \Delta V/\Delta d$ gives a better value for E when the value for Δd is small.

Q21.17. Reason: Because the post and leaf are both metals and are in contact with one another, they are at the same potential. The potential is the same everywhere inside and on the surface of a piece of metal, and the post and leaf are like a single piece of metal. If the potential were different on the leaf than on the post, there would be an electric field pointing from the leaf, to the post or from the post to the leaf, but we know that in electrostatics problems, the electric field in a metal is zero.
Assess: The nice things about metals in an electrostatics problem is that they have only one potential throughout. Even though the charge density on the metal and the electric field just outside the metal may be different at different places on the metal's surface, you can count on the potential to be constant at all points inside and on the metal.

Q21.19. Reason: (a) Since the capacitor remains connected to the battery, ΔV_C does not change and it is always equal to the potential difference of the battery $\Delta V_C = \Delta V_{\text{bat}}$.
(b) The capacitance of the capacitor depends on its construction ($C = \varepsilon_o A/d$). If the plate separation doubles, the capacitance will decrease by a factor of two.
(c) The capacitance of a capacitor is a measure of the amount of charge that can be placed on the capacitor per unit of electric potential placed across the plates of the capacitor ($C = Q/V$). If the capacitance decreases by a factor of two, then the amount of charge that a particular battery (fixed V) can put on the plates is reduced by a factor of two.
Assess: Since the battery remains connected, the potential difference across the capacitor remains constant but the charges may move around.

Q21.21. Reason: The electric potential at a distance r from a source charge Q may be determined by

$$V = kQ/r = (9.0 \times 10^9 \text{ N} \cdot \text{m}^2/\text{C}^2)(1.0 \times 10^{-9} \text{ C})/(1.0 \text{ m}) = 9.0 \text{ V}$$

The correct choice is A.

Assess: As symmetry would dictate, this is true at any point 1.0 m from charge A. The angle is completely irrelevant, as it could have been measured from any radial line.

Q21.25.   Reason: Knowing the difference in electric potential ΔV and the change in position d for two equipotential lines, we can determine the electric field near the midpoint between these two equipotential lines by $E = \Delta V/d$, (if the field is nonuniform, the smaller d, the more accurate the value of E). Using the equipotential lines on either side of the point of interest, we get for each case that $\Delta V_a = \Delta V_b = \Delta V_c = 100$ V.

We can do an eyeball comparison of each d (the separation between the two equipotential lines on either side of the point of interest) to get $d_a > d_b > d_c$, which allows us to write $E_c > E_b > E_a$. The correct choice is C.

Assess: This question emphasizes the relationship between electric potential difference and electric field.

Q21.29.   Reason: Use the relationship between the electric potential across the zapper, the separation between the plates, and the electric field to obtain the plate separation:

$$d = \Delta V/E = 4.0\times10^3 \text{ V}/(1.0\times10^6 \text{ V/m}) = 4.0\times10^{-3} \text{ m} = 0.40 \text{ cm}$$

Actually ΔV needs to be a little less than the breakdown value used; this would give a slightly larger value for the separation than calculated here. This is close to the answer of 0.5 cm. The correct choice is B.

Assess: Anything less than a few cm is reasonable. The calculated separation will only work for bugs less than a half centimeter in size. Knowing how the zapper works, it is somewhat ironic that the bug actually participates in its demise.

Q21.31.   Reason: You can see from the following figure that points 1 and 2 have negative potential and point 3 has positive potential. This makes $V_1 - V_3$ negative and so rules out B. The figure also shows that the potential is more negative at 2 than at 1, ruling out $V_2 - V_1$, choice C. Now both A and D are positive but it is hard to tell which one is larger. We reason that since point 2 is on a very small equipotential curve close to the positive charge, the potential at 2 is a very large negative number, and so $V_1 - V_2 = |V_2| - |V_1|$, in which that large negative number is subtracted, is probably larger than $V_3 - V_1 = |V_3| + |V_1|$, in which both of the potentials are only moderately large. The correct choice is A.

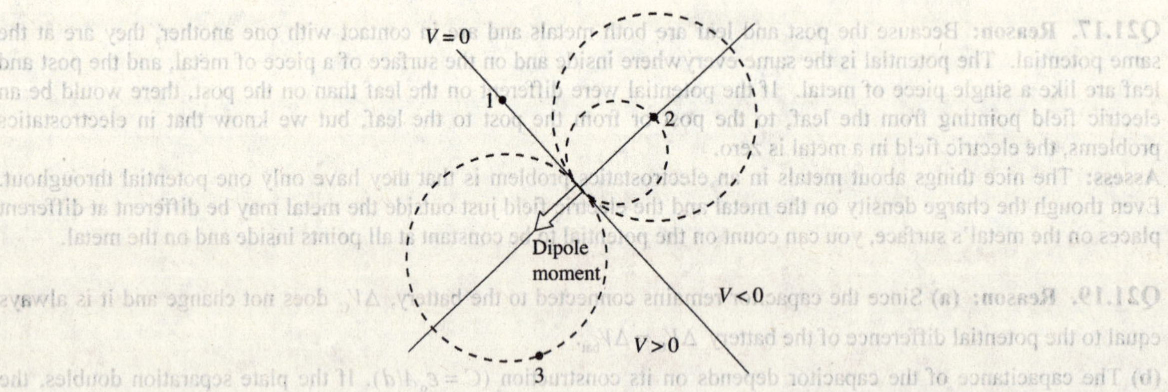

Assess: Often drawing a detailed picture will help us find the answer to a problem. Here, sketching the equipotentials has given us a deeper insight, namely, that $|V_2| > |V_1|$ and $|V_2| > V_3$.

Problems

P21.1.   Prepare: The work done is equal to the change in electric potential energy: $W = \Delta U_{elec} = q\Delta V$.
We are given that $\Delta V = 150 \text{ V} - 300 \text{ V} = -150 \text{ V}$.

Solve: Solve the first equation for q:

$$q = \frac{\Delta U_{elec}}{\Delta V} = \frac{4.5 \times 10^{-4}\,\text{J}}{-150\,\text{V}} = -3.0 \times 10^{-6}\,\text{C}$$

Assess: The answer is negative because it requires positive work to move a negative charge to a lower potential. As for units, $\text{J/V} = \text{C}$.

P21.3. Prepare: The work done is equal to the change in electric potential energy: $W = \Delta U_{elec} = q\Delta V$. Solve the equation for ΔV: $\Delta V = \Delta U_{elec}/q$.

Solve:

As the charge is moved from A to B, its electric potential changes by

$$\Delta V_{A \to B} = \frac{(\Delta U_{elec})_{A \to B}}{q} = \frac{3.0\,\mu\text{J}}{15\,\text{nC}} = 200\,\text{V}$$

Thus B is at a higher potential than A.

When the charges is moved from C to B, its electric potential changes by

$$\Delta V_{C \to B} = \frac{(\Delta U_{elec})_{C \to B}}{q} = \frac{-5.0\,\mu\text{J}}{15\,\text{nC}} = -330\,\text{V}$$

This means B is at a lower potential than C. Thus if the charge were moved in the *opposite* direction, from B to C, its electric potential would *increase*, so that $\Delta V_{B \to C} = -\Delta V_{C \to B} = +330\,\text{V}$. The total change in electric potential in moving from A to C is then $V_C - V_A = \Delta V_{AC} = \Delta V_{AB} + \Delta V_{BC} = 200\,\text{V} - (-330\,\text{V}) = 530\,\text{V}$.

Assess: The potential at B is between the potential at A and the potential at C. So if we move the charge from A to B and then to C, it will take a total work of $3.0\,\mu\text{J} - (-5.0\,\mu\text{J}) = 8.0\,\mu\text{J}$.

$$\Delta V_{AC} = \frac{(\Delta U_{elec})_{AC}}{q} = \frac{8.0\,\mu\text{J}}{15\,\text{nC}} = 530\,\text{V}$$

P21.9. Prepare: Energy is conserved. The potential energy is determined by the electric potential. The figure shows a before-and-after pictorial representation of an electron moving through a potential difference.

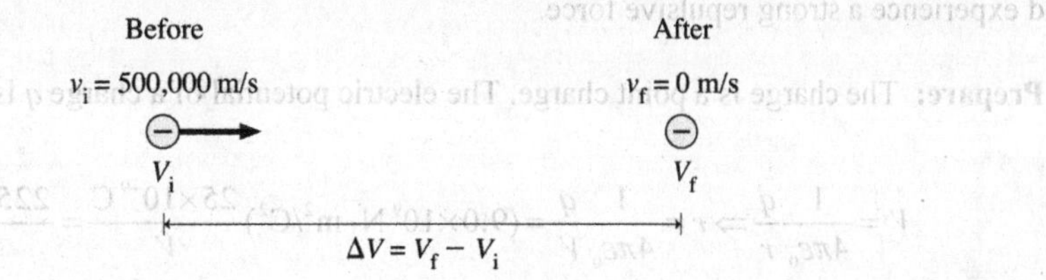

Solve: (a) Because the electron is a negative charge and it slows down as it travels, it must be moving from a region of higher potential to a region of lower potential.

(b) Using the conservation of energy equation,

$$K_f + U_f = K_i + U_i \Rightarrow K_f + qV_f = K_i + qV_i \Rightarrow V_f - V_i = \frac{1}{q}(K_i - K_f) = \frac{1}{(-e)}\left(\frac{1}{2}mv_i^2 - 0\,\text{J}\right)$$

$$\Rightarrow \Delta V = -\frac{mv_i^2}{2e} = -\frac{(9.11 \times 10^{-31}\,\text{kg})(5.0 \times 10^5\,\text{m/s})^2}{2(1.60 \times 10^{-19}\,\text{C})} = -0.71\,\text{V}$$

(c)

$$K_i = \frac{1}{2}mv_i^2 = q\Delta V = (-e)(-0.712\,\text{V}) = 0.71\,\text{eV}$$

Assess: The negative sign with ΔV verifies that the electron moves from a higher potential region to a lower potential region.

P21.11. Prepare: Let the distance between the two charges be called $2r$. Then the distance from the observation point midway between the charges to either charge is r. The potential near a charged particle is given by Equation 21.11: $V = KQ/r$.

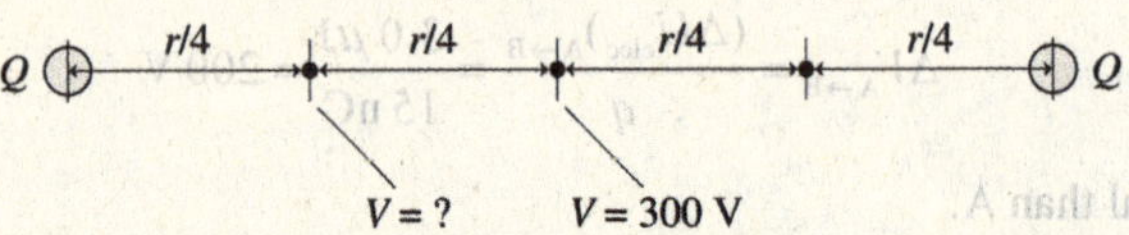

Solve: The total potential midway between the charges is equal to the sum of the potential of either charge by itself. Thus we have:

$$300\,\text{V} = \frac{KQ}{r} + \frac{KQ}{r} = \frac{2KQ}{r} \Rightarrow \frac{KQ}{r} = 150\,\text{V}$$

We don't know Q or r but this ratio of KQ and r is all we need to solve the problem. Since the total distance between the charges is $2r$, a point 25% of the way from one particle to the other is a distance $r/2$ from the closer particle and a distance $3r/2$ from the farther particle. Thus the potential at such a point is:

$$V = \frac{KQ}{r/2} + \frac{KQ}{3r/2} = \frac{8KQ}{3r} = \left(\frac{8}{3}\right)(150\,\text{V}) = 400\,\text{V}$$

Assess: Going from the midpoint of the two charges to a point closer to one of them increases the potential. This means that we would have to do work to move a positive test charge from the midpoint toward one of the charges. This makes sense considering that very close to either charge, the field is strong and a positive test charge placed there would experience a strong repulsive force.

P21.15. Prepare: The charge is a point charge. The electric potential of a charge q is given by Equation 21.10.
Solve:

$$V = \frac{1}{4\pi\varepsilon_0}\frac{q}{r} \Rightarrow r = \frac{1}{4\pi\varepsilon_0}\frac{q}{V} = (9.0\times10^9\,\text{N}\cdot\text{m}^2/\text{C}^2)\frac{25\times10^{-9}\,\text{C}}{V} = \frac{225\,\text{N}\cdot\text{m}^2/\text{C}}{V}$$

For $V = 2000$ V,

$$r_{2000} = \frac{225\,\text{N}\cdot\text{m}^2/\text{C}}{2000\,\text{V}} = 0.113\,\text{m}$$

For $V = 3000$ V, $r_{3000} = 0.075$ m. So the 2000 V equipotential surface is $0.113\,\text{m} - 0.075\,\text{m} = 0.038\,\text{m}$ farther from the charge than the 3000 V equipotential survace is.

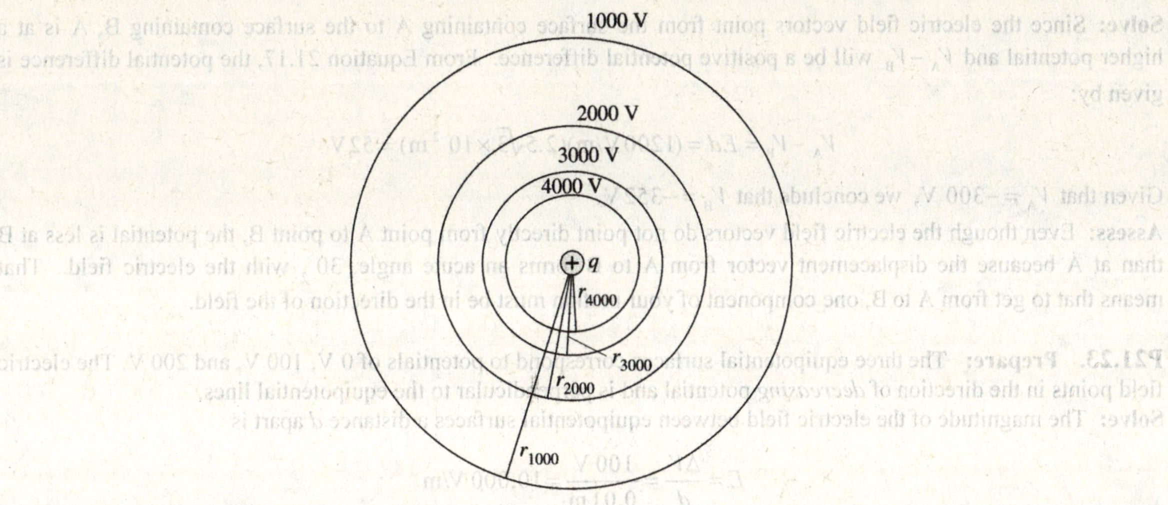

Assess: The radius of an equipotential surface increases with the decrease in potential. This is what we would expect because $V \propto 1/r$.

P21.17. Prepare: At the outer edge of a charged sphere of radius R, the electric potential is identical to that of a point charge Q at the center. That is,

$$V = \frac{1}{4\pi\varepsilon_0} = \frac{Q}{r} \quad r \geq R$$

Solve: The potential of the sphere is the potential right on the surface of the sphere. Thus,

$$Q = \frac{Vr}{1/4\pi\varepsilon_0} = \frac{(3400 \text{ V})(0.010 \text{ m})}{9.0\times10^9 \text{ N}\cdot\text{m}^2/\text{C}^2} = 3.8 \text{ nC}$$

P21.21. Prepare: In order to use Equation 21.17, $E = \Delta V / d$, we need to know the shortest distance between two equipotential surfaces. We are given the distance between point A which is on one surface and point B which is on another surface. However this is not the shortest distance, as you can see from the figure. The shortest distance, d, is given by $d = (5.0 \text{ cm}) \cos 30° = 2.5\sqrt{3}$ cm.

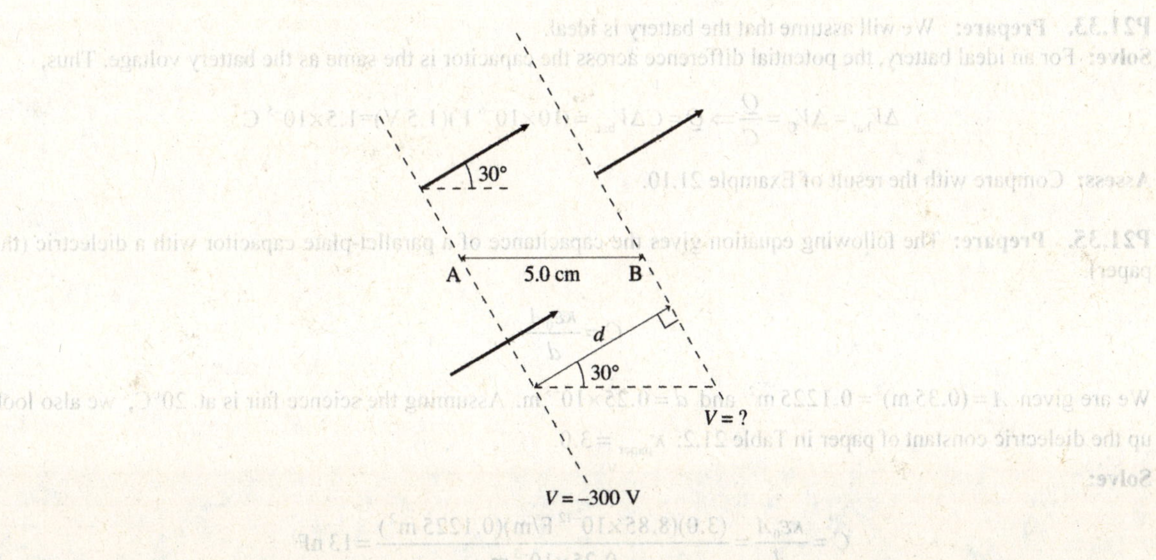

Solve: Since the electric field vectors point from the surface containing A to the surface containing B, A is at a higher potential and $V_A - V_B$ will be a positive potential difference. From Equation 21.17, the potential difference is given by:

$$V_A - V_B = Ed = (1200\,\text{V/m})(2.5\sqrt{3}\times10^{-2}\,\text{m}) = 52\,\text{V}$$

Given that $V_A = -300$ V, we conclude that $V_B = -352$ V.

Assess: Even though the electric field vectors do not point directly from point A to point B, the potential is less at B than at A because the displacement vector from A to B forms an acute angle, $30°$, with the electric field. That means that to get from A to B, one component of your motion must be in the direction of the field.

P21.23. Prepare: The three equipotential surfaces correspond to potentials of 0 V, 100 V, and 200 V. The electric field points in the direction of *decreasing* potential and is perpendicular to the equipotential lines.
Solve: The magnitude of the electric field between equipotential surfaces a distance d apart is

$$E = \frac{\Delta V}{d} = \frac{100\,\text{V}}{0.01\,\text{m}} = 10,000\,\text{V/m}$$

The direction of the electric field vector is "downhill," perpendicular to the equipotential surfaces, and directed to the left. That is, the electric field is 10,000 V/m to the left.

P21.25. Prepare: Examine Figure 21.29. In parts **(a)** and **(c)** of the problem electrode 3 is at a positive potential and electrode 1 is at a negative potential, so the difference $\Delta V_{31} = V_3 - V_1$ is positive.

In part **(b)** electrode 1 is at a positive potential and electrode 3 is at a slightly negative potential, so ΔV_{31} is negative.
Solve: (a) positive
(b) negative
(c) positive
Assess: Think of the positive potentials as up on a hill and negative potentials as in a valley.

P21.29. Prepare: We are thinking of a two-plate parallel capacitor. We will use Equation 21.20 with $A = L \times L$.
Solve: From Equation 21.20, the capacitance is

$$C = \frac{\varepsilon_0 A}{d} = \frac{\varepsilon_0 L^2}{d} \Rightarrow L = \sqrt{\frac{Cd}{\varepsilon_0}} = \sqrt{\frac{(100\times10^{-12}\,\text{F})(0.20\times10^{-3}\,\text{m})}{8.85\times10^{-12}\,\text{C}^2/\text{N}\cdot\text{m}^2}} = 0.0475\,\text{m} = 4.8\,\text{cm}$$

Assess: The capacitance depends only on the geometry of the capacitor.

P21.33. Prepare: We will assume that the battery is ideal.
Solve: For an ideal battery, the potential difference across the capacitor is the same as the battery voltage. Thus,

$$\Delta V_{\text{bat}} = \Delta V_C = \frac{Q}{C} \Rightarrow Q = C\Delta V_{\text{bat}} = (10\times10^{-6}\,\text{F})(1.5\,\text{V}) = 1.5\times10^{-5}\,\text{C}$$

Assess: Compare with the result of Example 21.10.

P21.35. Prepare: The following equation gives the capacitance of a parallel-plate capacitor with a dielectric (the paper).

$$C = \frac{\kappa\varepsilon_0 A}{d}$$

We are given $A = (0.35\,\text{m})^2 = 0.1225\,\text{m}^2$ and $d = 0.25\times10^{-3}\,\text{m}$. Assuming the science fair is at $20°\text{C}$, we also look up the dielectric constant of paper in Table 21.2: $\kappa_{\text{paper}} = 3.0$.
Solve:

$$C = \frac{\kappa\varepsilon_0 A}{d} = \frac{(3.0)(8.85\times10^{-12}\,\text{F/m})(0.1225\,\text{m}^2)}{0.25\times10^{-3}\,\text{m}} = 13\,\text{nF}$$

Assess: The answer could be expressed in scientific notation as 1.3×10^{-8} F but capacitances are usually given in pF (sometimes pronounced "puff"), nF, or μF. In our calculation the m^2 cancels, leaving F.

P21.37. **Prepare:** The battery establishes a potential difference of 9.0 V across the capacitor and this value persists with or without the Teflon. As for the electric field, we can use $E = \Delta V / d$ and since this equation does not refer to a dielectric constant, the electric field will be the same in both parts. We need to find the capacitance of the capacitor, both with and without the Teflon. Then we can use $Q = C\Delta V$ for the charge deposited.

Solve: **(a)** With the Teflon, the capacitance is given by:

$$C = \frac{\kappa \varepsilon_0 A}{d} = \frac{(2.0)(8.85 \times 10^{-12} \, C^2/(N \cdot m^2))\pi(1.0 \times 10^{-2} \, m)^2}{1.0 \times 10^{-4} \, m} = 55.6 \text{ pF}$$

The charge on the capacitor is given by $Q = C\Delta V = (55.6 \times 10^{-12} \text{ F})(9.0 \text{ V}) = 5.0 \times 10^{-10}$ C. The potential difference is just the EMF of the battery, 9.0 V and the electric field is given by $E = \Delta V / d = (9.0 \text{ V})/(1.0 \times 10^{-4} \text{ m}) = 9.0 \times 10^4$ V/m.

(b) After the Teflon is removed, the capacitance is reduced to

$$C = \frac{\varepsilon_0 A}{d} = \frac{(8.85 \times 10^{-12} \, C^2/(N \cdot m^2))\pi(1.0 \times 10^{-2} \, m)^2}{1.0 \times 10^{-4} \, m} = 27.8 \text{ pF}$$

Now the charge on the capacitor is $Q = C\Delta V = (27.8 \times 10^{-12} \text{ F})(9.0 \text{ V}) = 2.5 \times 10^{-10}$ C. The potential difference and electric field are what they were before, 9.0 V and 9.0×10^4 V/m, respectively.

Assess: Since the battery remained connected, the potential difference stayed the same even after the dielectric was removed. The electric field, which depends only on the potential difference and plate separation, was also unaffected. However, removing the dielectric caused some charge to flow back into the battery.

P21.39. **Prepare:** Assume the capacitor starts out with air between the plates. Since the battery is disconnected the charge on the capacitor remains the same. The capacitance of a capacitor may be determined by $C = Q/\Delta V$. When the capacitor is filled with a dielectric of dielectric constant κ, the capacitance increases to $C_\kappa = \kappa C$. Table 21.3 gives the dielectric constant for Teflon as 2.0.

Solve: **(a)** While the battery is connected the potential difference across the capacitor is 12.0 V. Simply disconnecting the battery does not change the potential difference across the capacitor; it retains its charge and nothing has changed yet. So the potential difference is still 12.0 V.

(b) After the Teflon is inserted the ratio of potential difference with teflon to air is calculated.

$$\frac{\Delta V_{teflon}}{\Delta V_{air}} = \frac{C_{teflon}/Q}{C_{air}/Q} = \frac{\kappa_{teflon}}{\kappa_{air}} = \frac{2.0}{1.0} = 2.0$$

The potential difference with the Teflon-filled capacitor is then $2.0(12.0 \text{ V}) = 24 \text{ V}$.

Assess: Since Pyrex glass has a greater dielectric constant, a capacitor filled with Pyrex glass will store more charge than one filled with paper when they are connected to the same electric potential difference (battery).

P21.43. **Prepare:** The potential difference for a capacitor may be determined by the capacitance and charge on the capacitor. If the spheres are initially neutral and you move a charge $+q$ from one sphere to the other, then one sphere has a charge of $+q$ and the other has a charge of $-q$. There is a charge of q on each sphere of the capacitor (one is positive and one is negative).

Solve: The final electric potential difference between the spheres is

$$V_F = \frac{q}{C} = \frac{12.0 \times 10^{-9} \, C}{24.0 \times 10^{-12} \, F} = 5.00 \times 10^2 \text{ V}$$

The work required to move the charge is equal to the energy stored in the charged capacitor:

$$W = QV/2 = (12.0 \times 10^{-9}\,\text{C})(5.00 \times 10^2\,\text{V})/2 = 3.00\,\mu\text{J}$$

Assess: These are small but not unreasonable values.

P21.45. Solve: (a) From Equations 21.22 and 21.20, the energy stored in the charged capacitor is

$$U_\text{C} = \frac{1}{2}C(\Delta V_\text{C})^2 = \frac{1}{2}\left(\frac{A\varepsilon_0}{d}\right)(\Delta V_\text{C})^2 = \frac{1}{2}\frac{\pi(0.01\,\text{m})^2(8.85 \times 10^{-12}\,\text{C}^2/(\text{N}\cdot\text{m}^2))}{0.50 \times 10^{-3}\,\text{m}}(200\,\text{V})^2 = 1.1 \times 10^{-7}\,\text{J}$$

(b) From Equation 21.25, the energy density in the electric field is

$$u_\text{E} = \frac{1}{2}\varepsilon_0 E^2 = \frac{1}{2}\varepsilon_0\left(\frac{\Delta V_\text{C}}{d}\right)^2 = \frac{1}{2}(8.85 \times 10^{-12}\,\text{C}^2/\text{N}\cdot\text{m}^2)\left(\frac{200\,\text{V}}{0.5 \times 10^{-3}\,\text{m}}\right)^2 = 0.71\,\text{J/m}^3$$

P21.49. Prepare: We can solve this problem by combining the formula for the electric field in the presence of a point charge: $E = \dfrac{KQ}{r^2}$, and the formula for the potential in the presence of a point charge: $V = \dfrac{KQ}{r}$.

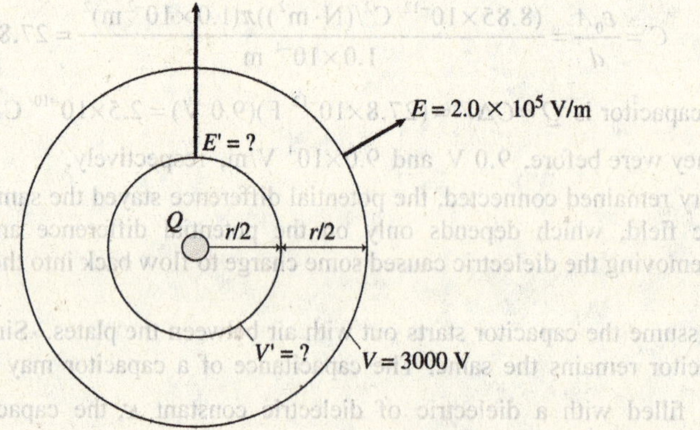

Solve: (a) If we divide the equation for V by the equation for E, we obtain $V/E = r$ so:

$$r = \frac{V}{E} = \frac{3000\,\text{V}}{2.0 \times 10^5\,\text{V/m}} = 1.5\,\text{cm}$$

(b) Since electric potential is inversely proportional to r, if we halve the distance to the charge, we double the potential. So at $r/2$, the potential is $V' = 6000$ V. The electric field on the other hand, is inversely proportional to r^2, so if we halve r, we quarter r^2 and quadruple the field: $E' = 4E$. Thus, at $r/2$, the electric field is $E' = 8.0 \times 10^5$ V/m.

Assess: To solve this problem, we didn't need to worry about the size of the charge because the ratio V/E is charge independent. This means that for any point charge, regardless of the size of the charge, the ratio of V to E at 1.5 cm equals 1.5 cm.

P21.51. Prepare: The charges are point charges. We will denote the protons with numbers 1 and 2 and the electron with number 3. The proton-electron distance can be obtained using Pythagorean theorem.
Solve: The electric potential energy of the electron, using Equation 21.9, is

$$U_\text{electron} = U_{13} + U_{23}$$

$$= (9.0 \times 10^9\,\text{N}\cdot\text{m}^2/\text{C}^2)\left[\frac{(1.60 \times 10^{-19}\,\text{C})(-1.60 \times 10^{-19}\,\text{C})}{\sqrt{(2.0 \times 10^{-9}\,\text{m})^2 + (0.5 \times 10^{-9}\,\text{m})^2}} + \frac{(1.60 \times 10^{-19}\,\text{C})(-1.60 \times 10^{-19}\,\text{C})}{\sqrt{(2.0 \times 10^{-9}\,\text{m})^2 + (0.5 \times 10^{-9}\,\text{m})^2}}\right]$$

$$= -1.12 \times 10^{-19}\,\text{J} - 1.12 \times 10^{-19}\,\text{J} = -2.24 \times 10^{-19}\,\text{J}$$

which will be reported as -2.2×10^{-19} J.

P21.55.   Prepare: The net potential is the sum of the scalar potentials due to each charge. Let the point on the y-axis where the electric potential is zero be at a distance y from the origin. At this point, $V_1 + V_2 = 0$ V.

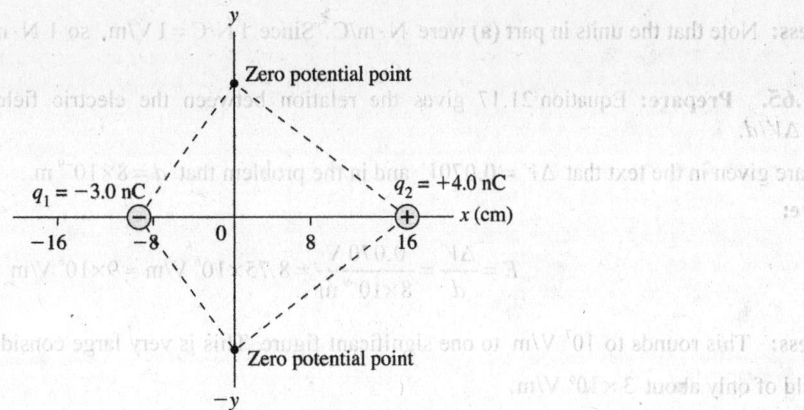

Solve: Using Equation 21.10,

$$\frac{1}{4\pi\varepsilon_0}\left\{\frac{q_1}{r_2}+\frac{q_2}{r_2}\right\}=0 \text{ V} \Rightarrow \frac{-3.0\times10^{-9}\,\text{C}}{\sqrt{(-9.0\,\text{cm})^2+y^2}}+\frac{4.0\times10^{-9}\,\text{C}}{\sqrt{(16.0\,\text{cm})^2+y^2}}=0$$

$$\Rightarrow 3\sqrt{(16\,\text{cm})^2+y^2}=4\sqrt{(-9\,\text{cm})^2+y^2} \Rightarrow 9(256\,\text{cm}^2+y^2)=16(81\,\text{cm}^2+y^2)$$

$$\Rightarrow 7y^2=1008\,\text{cm}^2 \Rightarrow y=\pm12\,\text{cm}.$$

Assess: In comparison to separation between the charges, these values seem reasonable.

P21.59.   Prepare: Outside a charged sphere the electric potential is identical to that of a point charge at the center and is given by Equation 21.11.
Solve: (a) For a proton, assumed to be a point charge, the electric potential is

$$V=\frac{1}{4\pi\varepsilon_0}\frac{(+e)}{r}=(8.99\times10^9\,\text{N}\cdot\text{m}^2/\text{C}^2)\frac{1.60\times10^{-19}\,\text{C}}{0.053\times10^{-9}\,\text{m}}=27.14\,\text{V}$$

which we report as 27 V to two significant figures.

(b) The potential energy of a charge q at a point where the potential is V is $U=qV$. The potential energy of the electron in the proton's potential is

$$U=(-1.60\times10^{-19}\,\text{C})\,(27.14\,\text{V})=-4.3\times10^{-18}\,\text{J}$$

P21.61.   Prepare: The net potential is the sum of the potentials due to each charge. Let $q_1=+5$ nC, $q_2=-5$ nC, and $q_3=10$ nC. Also, $r_1=2$ cm, $r_2=4$ cm, and $r_3=\sqrt{(2\,\text{cm})^2+(4\,\text{cm})^2}=4.47$ cm.
Solve: (a) Potential is a scalar, not a vector, so the net potential is simply the sum of the potentials of each of the charges. Each individual potential is simply that of a point charge, so

$$V_{\text{A}}=V_1+V_2+V_3=\frac{q_1}{4\pi\varepsilon_0 r_1}+\frac{q_2}{4\pi\varepsilon_0 r_2}+\frac{q_3}{4\pi\varepsilon_0 r_3}=\frac{1}{4\pi\varepsilon_0}\left(\frac{q_1}{r_1}+\frac{q_2}{r_2}+\frac{q_3}{r_3}\right)$$

$$=(9.0\times10^9\,\text{N}\cdot\text{m}^2/\text{C}^2)\left(\frac{5\times10^{-9}\,\text{C}}{0.02\,\text{m}}+\frac{-5\times10^{-9}\,\text{C}}{0.04\,\text{m}}+\frac{10\times10^{-9}\,\text{C}}{0.0447\,\text{m}}\right)=3140\,\text{V}=3100\,\text{V}$$

(b) The potential energy of a proton at point A is

$$U_{proton} = q_{proton}V_A = eV_A = (1.6\times10^{-19}\ \text{C})(3140\ \text{V}) = 5.0\times10^{-16}\ \text{J}$$

Assess: Note that the units in part **(a)** were $\text{N}\cdot\text{m/C}$. Since $1\ \text{N/C} = 1\ \text{V/m}$, so $1\ \text{N}\cdot\text{m/C} = 1\ \text{V}\cdot\text{m/m} = 1\ \text{V}$.

P21.65.   Prepare: Equation 21.17 gives the relation between the electric field and the potential difference: $E = \Delta V/d$.

We are given in the text that $\Delta V = 0.070V$ and in the problem that $d = 8\times10^{-9}\ \text{m}$.
Solve:

$$E = \frac{\Delta V}{d} = \frac{0.070\ \text{V}}{8\times10^{-9}\ \text{m}} = 8.75\times10^{6}\ \text{V/m} \approx 9\times10^{6}\ \text{V/m}$$

Assess: This rounds to $10^7\ \text{V/m}$ to one significant figure. This is very large considering that a spark in air requires a field of only about $3\times10^{6}\ \text{V/m}$.

P21.67.   Prepare: If there is an electric potential difference ΔV across a cell membrane of thickness d, the electric field in the membrane is $E = \Delta V/d$. This electric field exerts a force on a charged particle of magnitude $F = qE = q\Delta V/d$.
Solve: The force on the molecular ion is

$$F = q\Delta V/d = (10e)V/d = (10)(1.6\times10^{-19}\ \text{C})(7\times10^{-2}\ \text{V})/5.0\times10^{-9}\ \text{m} = 2.2\times10^{-11}\ \text{N}$$

Assess: We expect the force to be small, because the electric field and charge are both small.

P21.69.   Prepare: The proton will be attracted to the upper, negative plate and repelled by the lower positive plate. Since it is fired in *halfway* between the plates, by the time it hits the upper plate it will have moved through a potential difference of $\Delta V = 2500\ \text{V}$, so it will have gained 2500 eV of kinetic energy.   We convert eV to J: $2500\ \text{eV} = 4.0\times10^{-16}\ \text{J}$.
Solve: Use an energy equation.

$$K_f = K_i + \Delta K = \tfrac{1}{2}mv_i^2 + \Delta K = \tfrac{1}{2}(1.67\times10^{-27}\ \text{kg})(3.0\times10^{5}\ \text{m/s})^2 + 4.0\times10^{-16}\ \text{J} = 4.75\times10^{-16}\ \text{J}$$

Now solve for the final speed.

$$v_f = \sqrt{\frac{2K_f}{m}} = \sqrt{\frac{2(4.75\times10^{-16}\ \text{J})}{1.67\times10^{-27}\ \text{kg}}} = 7.5\times10^{5}\ \text{m/s}$$

Assess: We expected the final speed to be greater than the initial speed, but in the same ballpark, and it is.

P21.73.   Prepare: The electron has charge $q = -e$, and its potential energy at a point where the capacitor's potential is V is $U = -eV$. Since the electron is launched from the negative (lower potential) plate toward the positive (higher potential) plate, its potential energy becomes more negative (because of the negative sign of the electron charge). That is, the potential energy decreases, which must lead to an increase in the kinetic energy. Conversely, the electron's speed as it is launched is smaller than $2.0\times10^{7}\ \text{m/s}$. Energy is conserved. The electron's potential energy inside the capacitor can be found from the capacitor's electric potential.
Solve: **(a)** From Equation 21.6, the voltage across the capacitor is

$$\Delta V_C = Ed = (5.0\times10^{5}\ \text{V/m})(2.0\times10^{-3}\ \text{m}) = 1000\ \text{V}$$

(b) Because $E = \Delta V_C/d$, $Q = C\Delta V_C$, and $C = \varepsilon_0 A/d$, so $E = Q/\varepsilon_0 A$. Thus, the charge on each plate is

$$Q = \pi R^2 E\varepsilon_0 = \pi(1.0\times10^{-2}\ \text{m})^2(5.0\times10^{5}\ \text{V/m})(8.85\times10^{-12}\ \text{C}^2/\text{N}\cdot\text{m}^2) = 1.4\times10^{-9}\ \text{C}$$

(c) The conservation of energy equation is

$$K_f + qV_f = K_i + qV_i \Rightarrow \frac{1}{2}mv_i^2 = \frac{1}{2}mv_f^2 + q(V_f - V_i) \Rightarrow v_i^2 = v_f^2 + \frac{2}{m}(-e)(1000 \text{ V})$$

$$\Rightarrow v_i = \sqrt{(2.0\times10^7 \text{ m/s})^2 - \frac{2(1.60\times10^{-19} \text{ C})(1000 \text{ V})}{9.11\times10^{-31} \text{ kg}}} = 7.0\times10^6 \text{ m/s}$$

P21.75. Prepare: As the electron moves toward the sphere, the only force acting on it is the Coulomb force, which is a conservative force. As a result, energy is conserved. At the point where the electron is stopped and reflected back, all its initial kinetic energy $(KE = mv^2/2)$ has been converted into electric potential energy $(U = keQ/r)$. Once we know where the electron is momentarily stopped, we can find the electric field at this point, the force on the electron at this point, and finally, the electron's acceleration at this point.

Solve: (a) Knowing that energy is conserved, we can write that the total initial energy is equal to the total final energy: $K_i + U_i = K_f + U_f$.

Since the electron is fired from far away, its initial electric potential energy is zero. Since the electron moves toward the sphere until it is momentarily stopped, the final kinetic energy is zero. This is written as $KE_i = U_f$.

Inserting expressions for the kinetic end electric potential energy obtain $mv^2/2 = keQ/r$. This may by solved for v to obtain $v = \sqrt{2keQ/(rm)}$.

Note that r in this expression is the radius of the sphere (1.25 mm) plus the distance the electron stops from the surface of the sphere (0.30 mm) or $r = 1.55$ mm. Inserting values obtain v.

$$v = \sqrt{2(9.0\times10^9 \text{ N}\cdot\text{m}^2/\text{C}^2)(1.6\times10^{-19} \text{ C})(4.5\times10^{-9} \text{ C})/((1.55\times10^{-3} \text{ m})(9\times10^{-31} \text{ kg}))} = 9.6\times10^7 \text{ m/s}$$

(b) When the speed of the electron is half its original value, its kinetic energy is one fourth its original value. This is because kinetic energy is proportional to the square of speed. Let us call the distance from the center of the charge to the electron at this half-original-speed point, r'. Since the total energy of the electron is equal to its initial kinetic energy, that is, $E_{total} = K_i$ and since its kinetic energy at the point in question is $1/4$ times its original value, we can use conservation of energy to say:

$$E_{total} = K_i = \frac{1}{4}K_i + U(r') \Rightarrow U(r') = \frac{3}{4}K_i = \frac{3}{4}U_f$$

Now using the formula for the potential energy we can solve for r':

$$U(r') = \frac{keQ}{r'} = \frac{3}{4}U_f = \frac{3}{4}\frac{keQ}{r} \Rightarrow r' = \frac{4r}{3} = \frac{4(1.55 \text{ mm})}{3} = 2.07 \text{ mm}$$

When the electron has this distance from the center of the charge, its distance from the surface of the charge is $2.07 \text{ mm} - 1.25 \text{ mm} = 0.82 \text{ mm}$.

(c) The magnitude of the electric field at the turning point may be obtained by $E = kQ/r^2$. The force on the electron at the turning point is obtained by $F = eE$. Finally, the acceleration of the electron at its turnaround point is

$$a = F/m = eE/m = e(kQ/r^2)/m = keQ/(mr^2)$$

Inserting values

$$a = (9.0\times10^9 \text{ N}\cdot\text{m}^2/\text{C}^2)(1.6\times10^{-19} \text{ C})(4.5\times10^{-9} \text{ C})/((9.0\times10^{-31} \text{ kg})(1.55\times10^{-3} \text{ m})^2) = 3.0\times10^{18} \text{ m/s}^2$$

Assess: The electron is fired at the sphere with a speed that is about one third the speed of light. If the speed were much greater we would have to consider relativistic effects. The position where the electron has half the speed should be larger than the turnaround point, and it is four times larger. Finally, the acceleration is very large, but this is acceptable since the mass of the electron is so small.

P21.81. Prepare: Equipotentials are drawn by connecting points of equal potential. The field strength will be found using Equation 21.17.

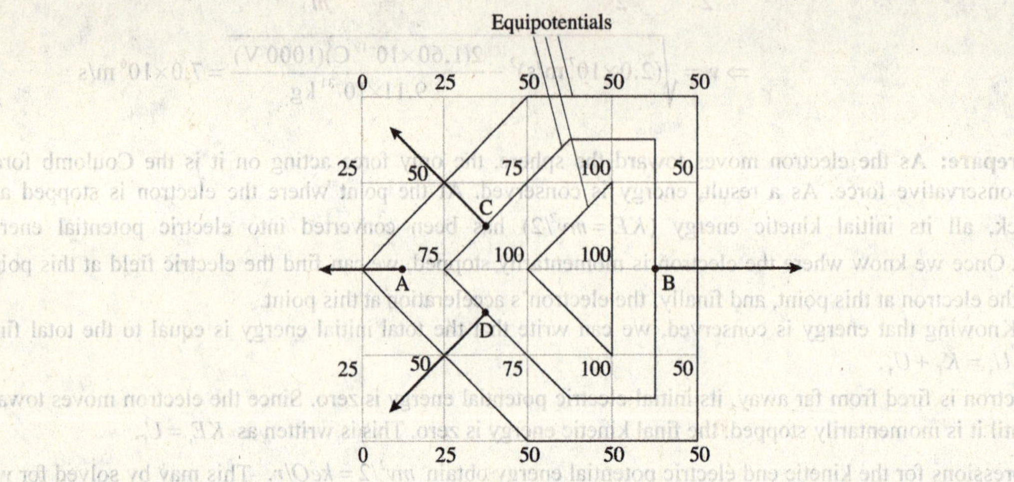

(b) To find $\vec{E}$, first determine the direction perpendicular to the equipotential and pointing toward a lower potential. The field strength is $E = \Delta V/d$, where d is the distance between equipotential lines that have potential difference ΔV. The electric field at point A points straight to the left, toward the 50 V equipotential. The distance between the 50 V line and the 75 V line is one grid spacing, or $d = 5$ cm. Thus,

$$E_A = \frac{25\ V}{0.05\ m} \approx 500\ V/m$$

As a vector, the field is $\vec{E}_A = (500\ V/m, left)$. Using the 50 V and 100 V equipotentials to calculate the field at point B, $E_B \approx (50\ V)/(0.05\ m) = 1000\ V/m$. Thus $\vec{E}_B = (1000\ V/m, right)$. The field at point C points at a 45° angle toward the 50 V equipotential. The distance d between the 50 V and 100 V equipotentials is $5\ cm \times \sqrt{2} = 7.07$ cm, so $E_C \approx (50\ V)/(0.0707\ m) = 710\ V/m$. As a vector, $\vec{E}_C = (710\ V/m,\ 45°$ above straight left). The field at point D is symmetrical with point C, so $\vec{E}_D = (710\ V/m,\ 45°$ below straight left).
(c) The electric field vectors are shown in the figure.

P21.83. Prepare: The energy stored in a capacitor is related to the physical size of the capacitor and the electric field in the capacitor by $U = \varepsilon A d E^2/2$.
Solve: (a) The energy stored in the air-filled capacitor may be determined by

$$U_{air} = \varepsilon_0 A d E^2/2 = (8.85\times10^{-12}\ C^2/(N\cdot m^2))(0.15\ m)^2(5.0\times10^{-4}\ m)(3.0\times10^6\ V/m)^2/2 = 4.5\times10^{-4}\ J$$

(b) The energy stored in the Teflon-filled capacitor may be determined by

$$U_{Teflon} = \varepsilon A d E^2/2 = (2\times8.85\times10^{-12}\ C^2/(N\cdot m^2))(0.15\ m)^2(5.0\times10^{-4}\ m)(60\times10^6\ V/m)^2/2 = 0.36\ J$$

Assess: The Teflon-filled capacitor can store 800 times the amount of energy as the air-filled capacitor.

22

CURRENT AND RESISTANCE

Q22.3. Reason: In order for electrons to move through a wire as a current, the wire must be connected to a source of electric potential difference. This source of electric potential difference will establish an electric field in the circuit and it is this electric field that causes electrons to travel through the circuit.

Assess: Frequently, we use misleading language when describing the current in a circuit. In fact, you may have used the misleading statement "flow of current." Current is defined as the motion of charges. It is the charges that flow, not the current. Current is the flow of charges.

Q22.5. Reason: The circuit has no place to store electrons and no way to generate electrons. Also, for every electron that enters the circuit, another electron leaves the circuit. Based on these two statements, we can conclude that I_1 is equal to I_4. As the circuit shows, the current I_1 splits into two currents (I_2 and I_3) which then combine to give I_4. This allows us to conclude that both I_1 and I_4 are greater than I_2 and I_3. Since the wires that carry the current I_2 and I_3 are identical (they each have the same resistance), the current I_1 will divide in half to give I_2 and I_3. This allows us to make the following statements:

$$I_2 = I_3 < I_1 = I_4, \quad \text{and} \quad (I_2 = I_3) = (1/2)(I_1 = I_4)$$

Assess: We cannot create or destroy charges in the wire, neither can we store them in the junction. The rate at which electrons flow into one or many wires must be exactly balanced by the rate at which they flow out of the other wires. We are conserving the amount of charge and the flow of that charge (current).

Q22.7. Reason: If negative charges move opposite the field, we still say it is a positive current in the direction of the field. So both the positive ions and the negative ions contribute to a current in the direction of the field.

Assess: The two ion currents do **not** cancel out, even if the ions were moving at the same speed.

Q22.11. Reason: The resistance of an object is given by Equation 22.7: $R = \rho L/A$.

For objects with a circular cross section, $A = \pi r^2$. Since all five wires are made of the same material, ρ is the same in each equation. Now we must compute the resistance of all five wires.

$$R_1 = \frac{\rho L}{\pi r^2}$$

$$R_2 = \frac{\rho L}{\pi (2r)^2} = \frac{1}{4}\frac{\rho L}{\pi r^2} = \frac{1}{4}R_1$$

$$R_3 = \frac{\rho (2L)}{\pi (2r)^2} = \frac{2}{4}\frac{\rho L}{\pi r^2} = \frac{1}{2}R_1$$

$$R_4 = \frac{\rho(2L)}{\pi r^2} = 2\,\frac{\rho L}{\pi r^2} = 2R_1$$

$$R_5 = \frac{\rho(4L)}{\pi(2r)^2} = \frac{4}{4}\,\frac{\rho L}{\pi r^2} = R_1$$

Therefore

$$R_4 > R_1 = R_5 > R_3 > R_2$$

Assess: This question helps cement the concept that longer wires have higher resistance while fatter wires have lower resistance.

Q22.13. Reason: Conservation of charge requires $I_1 = I_2 = I_3$ and $I_4 = I_7$. Symmetry requires $I_5 = I_6$. Conservation of charge requires $I_4 = I_5 + I_6$, so $I_4 > I_5$.

So far we know $I_4 = I_7 > I_5 = I_6$. Now we compare the two circuits. The potential difference from 1 to 3 to 2 is the same as the potential difference from 4 to 6 to 7 and the resistance of the wires around each path is the same, so $I = \Delta V/R$ is the same for both of these paths. So $I_3 = I_6$.

Putting this all together, $I_4 = I_7 > I_1 = I_2 = I_3 = I_5 = I_6$.

Assess: The stated assumptions (identical batteries, wires of equal radius) are important. Conservation of charge, symmetry, and Ohm's law are all helpful here.

Q22.15. Reason: Equation 22.8 gives the current in an ohmic wire: $I = \Delta V/R$.

We now compute the current in each wire.

$$I_1 = \frac{2\ \text{V}}{2\ \Omega} = 1\ \text{A}$$

$$I_2 = \frac{1\ \text{V}}{2\ \Omega} = \frac{1}{2}\ \text{A}$$

$$I_3 = \frac{2\ \text{V}}{1\ \Omega} = 2\ \text{A}$$

$$I_4 = \frac{1\ \text{V}}{1\ \Omega} = 1\ \text{A}$$

Therefore

$$I_3 > I_1 = I_4 > I_2$$

Assess: This exercise helps one see that potential difference causes the current while resistance, well, resists it.

Q22.21. Reason: The brightness is determined by the power dissipated in each bulb. When they both operate at 120 V we use the power equation $P = \dfrac{(\Delta V)^2}{R}$ to see that the 100 W bulb has a lower resistance than the 60 W bulb.

But when they both have the same current through them we use the other version of the power equation $P = I^2 R$ to show that the bulb with the lower resistance (the 100 W bulb) dissipates less power and is dimmer than the 60 W bulb.

Assess: The wattage label on a regular household bulb assumes it will be operated at 120 V; the power dissipated will be different if the potential difference is not 120 V.

Q22.23. Reason: We know the power dissipated by a resistor is a product of the voltage (potential difference) and the current, $P_R = I\Delta V_R$. For objects that obey Ohm's law, this can also be written as Equation 22.13. While not strictly true, we will assume the light bulb filaments in this question obey Ohm's law so we can apply this relation: $P_R = (\Delta V_R)^2/R$.

Solve for R and apply to each case.

A.

$$R = \frac{(\Delta V_R)^2}{P_R} = \frac{(1.5\text{ V})^2}{0.8\text{ W}} = 2.8\ \Omega$$

B.

$$R = \frac{(\Delta V_R)^2}{P_R} = \frac{(3\text{ V})^2}{6\text{ W}} = 1.5\ \Omega$$

C.

$$R = \frac{(\Delta V_R)^2}{P_R} = \frac{(4.5\text{ V})^2}{4\text{ W}} = 5.1\ \Omega$$

D.

$$R = \frac{(\Delta V_R)^2}{P_R} = \frac{(6\text{ V})^2}{8\text{ W}} = 4.5\ \Omega$$

The correct choice is C.

Assess: The fact that ΔV_R is squared carries the day for C.

Q22.25. Reason: If the potential difference is increased then the current will also increase, and so will the power dissipated $(P_R = I\Delta V_R)$. The length is fixed, so the electric field will also increase $(E = \Delta V/d)$.

Resistance is a function of the wire itself (the material it is made of and the temperature), but not directly a function of the potential difference.

The correct choice is C.

Assess: Indirectly, since the increased potential difference increases the current and power dissipation, the temperature may rise and cause the resistance to change too, but we don't know whether the temperature coefficient is positive or negative (the resistance might go up or down).

Q22.27. Reason: Use $P = \dfrac{(\Delta V)^2}{R}$ for the power dissipated. Since the resistor is the same, R doesn't change, and we see that the power scales as the square of the potential difference. Doubling the potential difference (from 3.0 V to 6.0 V) changes the power dissipated by a factor of 4. $(4)(1.0\text{ W}) = 4.0\text{ W}$. The answer is D.

Assess: This means the current increases also.

Problems

P22.3. Prepare: By conservation of charge, the sum of the currents into a junction must equal the sum of the currents leaving the junction.

Solve: (a) Therefore, we know that the current in wire 3 must be $0.65\text{ A} - 0.40\text{ A} = 0.25\text{ A}$. This is 0.25 C per second.

$$\frac{0.25\text{ C}}{1\text{ s}}\left(\frac{1\ e}{1.6\times10^{-19}\text{ C}}\right) = 1.6\times10^{18}\text{ electrons/s}$$

(b) Because more current is coming out of the junction in wire 2 than is going in in wire 1, the 0.25 A in wire 3 must be going into the junction.

Assess: Electrons can't disappear, so each one that comes into the junction must leave it.

P22.5. Prepare: Electric current, charge, and time are related by $I = \Delta Q/\Delta t$. The amount of charge due to N electrons is $\Delta q = Ne$.

Solve: The number of electrons passing a given point in 1.0 s may be determined by

$$N = \Delta Q/e = I\Delta t/e = (15\times10^{-6}\,\text{A})(1.0\,\text{s})/(1.6\times10^{-19}\,\text{C}) = 9.4\times10^{13} \text{ electrons}$$

Assess: Since each electron has such a small charge, we expect a large number of electrons.

P22.7. Prepare: Current is defined in terms of the amount of charge ΔQ passing through a cross section of the wire in a time interval Δt: $I = \Delta Q/\Delta t$.
We are given that $\Delta t = 40\,\mu\text{s} = 40\times10^{-6}\,\text{s}$ and $\Delta Q = (1-0.13)(6.0\times10^{-4}\,\text{C}) = (0.87)(6.0\times10^{-4}\,\text{C}) = 5.22\times10^{-4}\,\text{C}$.
Solve:

$$I = \frac{\Delta Q}{\Delta t} = \frac{5.22\times10^{-4}\,\text{C}}{40\times10^{-6}\,\text{s}} = 13\,\text{A}$$

Assess: 13 A is a fairly large current, but that is because capacitors discharge very quickly when the plates are connected with a wire.
In the next chapter you will learn more specifically how the current decays in a discharging capacitor; it isn't at a constant rate or in a linear fashion. But since all we're asked for in this problem is the average current, we are already equipped to solve the problem.

P22.11. Prepare: We believe in conservation of current at a junction, expressed in Equation 22.3: $\Sigma I_{\text{in}} = \Sigma I_{\text{out}}$. We apply this first at the left junction (because there is only one unknown current there), and then at the right.
Solve: At the left junction:

$$\Sigma I_{\text{in}} = \Sigma I_{\text{out}}$$
$$7\,\text{A} + I_C = 5\,\text{A}$$

So $I_C = 5\,\text{A} - 7\,\text{A} = -2\,\text{A}$, where the negative sign indicates the current goes in the opposite direction from the arrow (or, to the right).
At the right junction:

$$\Sigma I_{\text{in}} = \Sigma I_{\text{out}}$$
$$3\,\text{A} = I_B + I_C$$
$$3\,\text{A} = I_B + (-2\,\text{A})$$

So $I_B = 3\,\text{A} + 2\,\text{A} = 5\,\text{A}$.

Assess: On the diagram, replace the unknown current labels with their newly known values. Use the negative sign on I_C to change the direction of the arrow. Verify that current is conserved at each junction and everything makes sense.

P22.15. Prepare: Charge, electric current, and time are related by $I = \Delta Q/\Delta t$. Work, electric potential difference, and charge are related by $W = \Delta q\Delta V$.

Solve: (a) The charge transferred is $\Delta q = I\Delta t = (2.5\times10^{-3}\,\text{A})(5.0\,\text{hr})(3.6\times10^{3}\,\text{s/hr}) = 45\,\text{C}$

(b) The change in potential energy of these charges by the battery is $\Delta U = \Delta q\Delta V = (45\,\text{C})(9.0\,\text{V}) = 4.1\times10^{2}\,\text{J}$.

Assess: These values are reasonable for this case.

P22.17. Reason: The total potential difference is related to the potential difference for one cell by $V_{\text{total}} = NV_{\text{cell}}$.
Solve: The number of cells needed to generate a total potential difference of 350 V is

$$N = V_{\text{total}}/V_{\text{cell}} = 350\,\text{V}/(0.110\,\text{V}) = 3200\,\text{cells}.$$

Assess: Since the potential difference per cell is small, we expect a large number of cells.

P22.19. Prepare: Resistance and resistivity are related through Equation 22.7. Resistivity depends only on the type of material and not on the geometry of the wire. Equation 22.7 can also be written as $R = \rho L/A = \rho L/(\pi r^2)$, where the wire has length L and radius r.
Solve: (a) Wires 1 and 2 are made of the same material, so $\rho_2 = \rho_1$ and thus $\rho_2/\rho_1 = 1.0$.
(b) Because the two wires have the same resistivity,

$$\frac{R_2}{R_1} = \frac{\rho L_2/(\pi r_2^2)}{\rho L_1/(\pi r_1^2)} = \left(\frac{r_1}{r_2}\right)^2 \frac{L_2}{L_1} = \left(\frac{1}{2}\right)^2 \frac{2}{1} = \frac{1}{2} = 0.50$$

Assess: A thicker wire has smaller resistance and a larger wire has higher resistance.

P22.21. Prepare: Resistivity is related to resistance, length, and cross-sectional area by $R = \rho L/A$, and the area is related to the width and thickness of the leaf by $A = WT$.
Solve: Combining these two expressions and solving for the resistivity, we obtain

$$\rho = RA/L = RWT/L = (2.0\times10^6\ \Omega)(2.5\times10^{-2}\ \text{m})(2.0\times10^{-4}\ \text{m})/(0.20\ \text{m}) = 50\ \Omega\cdot\text{m}$$

Assess: This value for the resistivity is the same order of magnitude as other organic materials listed in Table 22.1.

P22.25. Prepare: The current I in a wire when a potential difference is applied to the ends of the wire can be obtained from Equation 22.5, $I = \Delta V/R$, and Equation 22.7, $R = \rho L/A$. The resistivity of nichrome from Table 22.1 is $1.5\times10^{-6}\ \Omega\cdot\text{m}$.
Solve:

$$I = \Delta V\left(\frac{A}{\rho L}\right) = \frac{(3.0\ \text{V})\pi(0.40\times10^{-3}\ \text{m})^2}{(1.5\times10^{-6}\ \Omega\cdot\text{m})(0.50\ \text{m})} = 2.0\ \text{A}$$

Assess: The resistivity of nichrome is small, so a current of 2.0 A is reasonable.

P22.29. Prepare: The slope of the I versus ΔV graph, according to Ohm's law, is the reciprocal of the resistance R. The resistance is directly proportional to the length of the resistor according to Equation 22.7.
Solve: (a) From the graph in the figure, the reciprocal of the slope and hence the resistance of the resistor is

$$\frac{\Delta V}{I} = R = \frac{1}{\text{Slope}} = \frac{10\ \text{V}}{5.0\ \text{A}} = 2.0\ \Omega$$

(b) Doubling the length of the wire will double the resistance, which means the slope will decrease. The current-versus-potential-difference graph is shown.

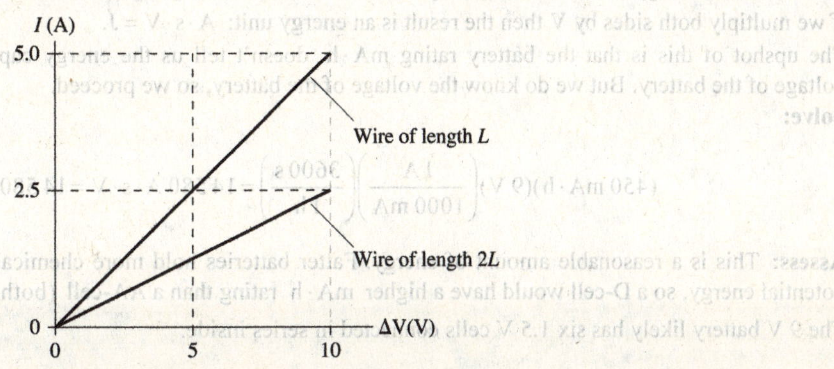

Assess: Since the resistance is the reciprocal of the slope, a smaller slope means a higher resistance.

P22.31. Prepare: We'll assume the filament is ohmic so we can use Ohm's law. Equation 22.7 will help us find the resistance, and we need to recall $E = \Delta V/d$.

Solve: (a) $E = \Delta V/d = 120$ V/(0.60 m) $= 200$ V/m.

(b) All else being equal, the new field strength would be $E = \Delta V/d' = 120$ V/(1.20 m) $= 100$ V/m.

(c) Assuming the filament is ohmic, the original current would be

$$I = \frac{\Delta V}{R} = \frac{120\ \text{V}}{240\ \Omega} = 0.50\ \text{A}$$

We are reminded that the current is proportional to the electric field, so halving the electric field (from 200 V/m to 100 V/m) means we also halve the current from 0.50 A to 0.25 A.

(d) We can use Ohm's law with the new current to calculate the new resistance.

$$R' = \frac{\Delta V}{I'} = \frac{120\ \text{V}}{0.25\ \text{A}} = 480\ \Omega$$

Assess: Another way to see the new resistance: Since $R = \rho L/A$, then doubling the length doubles the resistance, and $R' = 2 \times 240\ \Omega = 480\ \Omega$. This method is consistent with the other one.

The power dissipated (the brightness of the bulb) will be reduced correspondingly because $P = (\Delta V)^2/R$. ΔV is still 120 V but R has doubled.

P22.33. Prepare: Assume the copper wire is a resistor. We'll need Ohm's law $\Delta V = IR$.

Solve: The text says the field in a current-carrying resistor is

$$E = \frac{\Delta V}{L} = \frac{IR}{L} = \frac{I(\rho \frac{L}{A})}{L} = \frac{I\rho}{A} = \frac{(20\ \text{A})(1.7 \times 10^{-8}\ \Omega \cdot \text{m})}{\pi(0.50\ \text{mm})^2} = 0.43\ \text{V/m}$$

Assess: This is not a large field strength.

P22.37. Prepare: The text explains that the potential energy gained is $\Delta U = (\Delta q)\mathcal{E}$.

Solve:

$$\mathcal{E} = \frac{\Delta U}{\Delta q} = \frac{7.2\ \text{J}}{1.2\ \text{C}} = 6.0\ \text{V}$$

Assess: This is a typical emf for a battery.

P22.41. Prepare: The dimensions of current times time (the rating printed on the battery) are not yet the dimensions of energy until we multiply by the dimensions of voltage. That is, in SI units, $A \cdot s = C$, not J. However, if we multiply both sides by V then the result is an energy unit: $A \cdot s \cdot V = J$.

The upshot of this is that the battery rating $mA \cdot h$ doesn't tell us the energy capacity unless we also know the voltage of the battery. But we do know the voltage of the battery, so we proceed.

Solve:

$$(450\ \text{mA} \cdot \text{h})(9\ \text{V})\left(\frac{1\ \text{A}}{1000\ \text{mA}}\right)\left(\frac{3600\ \text{s}}{1\ \text{h}}\right) = 14\,580\ \text{A} \cdot \text{s} \cdot \text{V} = 14\,580\ \text{J} \approx 15\,000\ \text{J}$$

Assess: This is a reasonable amount of energy. Fatter batteries hold more chemicals and therefore more chemical potential energy, so a D-cell would have a higher $mA \cdot h$ rating than a AA-cell (both at 1.5 V).

The 9 V battery likely has six 1.5 V cells connected in series inside.

P22.43.   **Prepare:** We are given the power and the voltage of the toaster, as well as the time interval.
Solve:

$$P = I\Delta V = \frac{\Delta Q}{\Delta t}\Delta V \Rightarrow \Delta Q = \frac{P\Delta t}{\Delta V} = \frac{(900\ \text{W})(60\ \text{s})}{120\ \text{V}} = 450\ \text{C}$$

Assess: These are typical values for a toaster.

P22.45.   **Prepare:** Knowing the area of the plate and the thickness of the desired plating, we can determine the volume of zinc to be plated (don't forget we have to plate both sides). Knowing the volume of zinc needed and the density of zinc, we can determine the mass of the zinc needed. Knowing the molecular mass of zinc and Avogadro's number, we can determine the mass of each zinc ion and then the number of ions needed to obtain the desired platting mass. Knowing the number of ions needed and the charge on each ion, we can determine the amount of charge to be transferred. Finally, knowing the amount of charge to be transferred and the current, we can determine the amount of time needed to transfer this charge.
Solve: The volume of zinc to be plated is

$$V = 2(\Delta x)A = 2(1.00\times10^{-7}\ \text{m})(4.0\times10^{-4}\ \text{m}^2) = 8.0\times10^{-11}\ \text{m}^3$$

The mass of zinc needed is

$$M = \rho V = (7.14\times10^3\ \text{kg/m}^3)(8.0\times10^{-11}\ \text{m}^3) = 5.71\times10^{-7}\ \text{kg}$$

The mass of one zinc ion is

$$m = M/N_A = (65.4\times10^{-3}\ \text{kg/mole})/(6.02\times10^{23}/\text{mole}) = 1.09\times10^{-25}\ \text{kg}$$

The number of zinc ions needed is

$$N = M/m = 5.71\times10^{-7}\ \text{kg}/(1.09\times10^{-25}\ \text{kg}) = 5.24\times10^{18}$$

The amount of charge carried by this number of zinc ions is

$$\Delta q = Nq_{\text{ion}} = N(2e) = (5.24\times10^{18})(2)(1.6\times10^{-19}\ \text{C}) = 1.68\ \text{C}$$

The amount of time needed to transport this amount of charge with the specified current is

$$\Delta t = \Delta q/I = 1.68\ \text{C}/(1.0\times10^{-3}\ \text{A}) = 1.68\times10^3\ \text{s} = 28\ \text{min}$$

Assess: This is a reasonable plating time. If you hang out with physics faculty long enough, you will eventually hear one mutter "sweet." When they do that, they have just thought up a problem similar to this one. Notice the large number of concepts involved and the scope and range of these concepts. This type problem is excellent for developing your logic skills. It also is excellent final exam material.

P22.49.   **Prepare:** The resistance is given by Equation 22.7: $R = \rho L/A$.
We are given $L = 5.0\times10^{-9}\ \text{m}$, $A = \pi r^2 = \pi(d/2)^2 = \pi(0.30\ \text{nm}/2)^2 = 7.07\times10^{-20}\ \text{m}^2$, and $\rho = 0.60\ \Omega\cdot\text{m}$.
Solve:

$$R = \frac{\rho L}{A} = \frac{(0.60\ \Omega\cdot\text{m})(5.0\times10^{-9}\ \text{m})}{7.07\times10^{-20}\ \text{m}^2} = 4.2\times10^{10}\ \Omega = 42\ \text{G}\Omega$$

Assess: This is a tremendously large resistance (due to the small cross-sectional area); in many electronic circuit applications it would be simply approximated as infinite. So not many charge carriers (ions) will get through the ion channel (the current will only be a few pA, depending on the potential difference); but that is precisely the point—we want the cell to control the ion flow very closely.

P22.51.   **Prepare:** The electric potential difference, current, and resistance are related by Ohm's law $I = \Delta V/R$.
The amount of charge, the current, and the time it takes the charge to flow are related by $\Delta Q = I\Delta t$.

Solve: The current is obtained by $I = \Delta V/R = 1.2\,\text{V}/22\,\Omega = 54.5\,\text{mA}$. The amount of time the battery can deliver this current is $\Delta t = \Delta Q/I = 1800\,\text{mA}\cdot\text{hr}/54.5\,\text{mA} = 33\,\text{h}$.

Assess: Since the current is very small (54.5 mA) we expect the battery to able to deliver this current for a long time.

P22.57. Prepare: The voltage drop along the wire is related to the resistance of the wire and the current in the wire by $\Delta V = IR$. The resistance of a wire is related to its resistivity, length, and cross-sectional area by $R = \rho L/A$. The area of a wire is related to its diameter by $A = \pi d^2/4$.

Solve: The resistance of wire is

$$R = \Delta V/I = 0.50\ \text{V}/(3\times10^2\ \text{A}) = 1.67\times10^{-3}\ \Omega$$

The cross-sectional area of the wire is

$$A = \rho L/R = (1.7\times10^{-8}\ \Omega\cdot\text{m})(1.0\ \text{m})/(1.67\times10^{-3}\ \Omega) = 1.02\times10^{-5}\ \text{m}^2$$

The diameter of the wire is

$$d = \sqrt{4A/\pi} = \sqrt{4(1.02\times10^{-5}\ \text{m}^2)/\pi} = 3.6\times10^{-3}\ \text{m} = 3.6\ \text{mm}$$

Assess: The next time you are working on your car, take a look at the cable between the battery and the starter. You will find that this is a good value.

P22.59. Prepare: current is conserved, the currents in the two segments of the wire are the same. The currents in the two segments of the wire are related by $I_1 = I_2 = I$. $\Delta V_1 = I_1 R_1$ and $\Delta V_2 = I_2 R_2$. Because from Equation 22.7 $R = \rho L/A$ so $R_1 = \rho_1 L_1/A_1$ and $R_2 = \rho_2 L_2/A_2$. The wire's diameter is constant, so $A_1 = A_2$. Also $L_1 = L_2$.

Solve: From this information,

$$\frac{\Delta V_1}{\Delta V_2} = \frac{I_1 R_1}{I_2 R_2} = \left(\frac{I_1}{I_2}\right)\left(\frac{\rho_1 L_1}{A_1}\right)\left(\frac{A_2}{\rho_2 L_2}\right) = \frac{\rho_1}{\rho_2} = \frac{1}{2}$$

Assess: Notice the beauty of being asked to find a ratio of two quantities—most other quantities cancel.

P22.61. Prepare: Since the problem asks us to identify the likely material by its electrical properties, we probably want to find the resistivity of the material. The equation that contains resistivity is $R = \rho L/A$.

Fortunately, we are also given L and A (indirectly), and we can compute R from Ohm's law, $R = \Delta V/I$. We'll compute ρ and then look up the value in Table 22.1 to see if we can identify the material.

We are given $L = 2.3\ \text{m}$, $\Delta V = 1.2\ \text{V}$, $I = 0.61\ \text{A}$, and $A = \pi r^2 = \pi(d/2)^2 = \pi(0.38\ \text{mm}/2)^2 = 1.13\times10^{-7}\ \text{m}^2$.

Solve: Solve the first equation for ρ and insert R from Ohm's law.

$$\rho = \frac{RA}{L} = \frac{\Delta V}{I}\frac{A}{L} = \frac{(1.2\ \text{V})(1.13\times10^{-7}\ \text{m}^2)}{(0.61\ \text{A})(2.3\ \text{m})} = 9.7\times10^{-8}\ \Omega\cdot\text{m}$$

This value matches the value given in the table for iron, so we assert the wire is made of iron.

Assess: We are gratified that the answer matched one of the values in the table. The units work too, as $\text{V/A} = \Omega$.

P22.65. Prepare: The power dissipated by the heater is related to the potential difference across the heater and its resistance by $P = \Delta V^2/R$. The amount of energy needed to raise a mass m of water by an amount ΔT is $Q = mc\Delta T$. The power dissipated is related to the energy dissipated and the time to dissipate that energy by $P = W/\Delta t$.

Solve: (a) The resistance of the heater is

$$R = \Delta V^2/P = (120\ \text{V})^2/(300\ \text{W}) = 48\ \Omega$$

(b) The amount of energy that must be dissipated by the heater in order to heat the water from $18°C$ to $100°C$ is

$$Q = mc\Delta T = (0.40\text{ kg})(4.186\times10^3\text{ J/(kg}\cdot°C))(82°C) = 1.37\times10^5\text{ J}$$

The time required to dissipate this much energy is

$$\Delta t = Q/P = 1.37\times10^5\text{ J}/(3\times10^2\text{ W}) = 460\text{ s} = 7.6\text{ min}$$

Assess: This is not an unreasonable amount of time to wait for a good cup of tea.

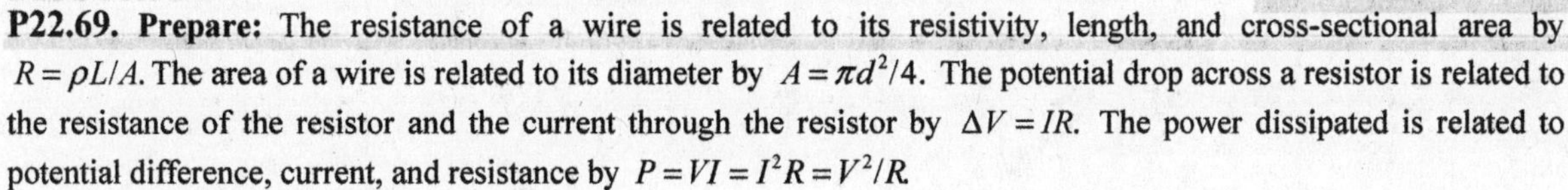

P22.69. Prepare: The resistance of a wire is related to its resistivity, length, and cross-sectional area by $R = \rho L/A$. The area of a wire is related to its diameter by $A = \pi d^2/4$. The potential drop across a resistor is related to the resistance of the resistor and the current through the resistor by $\Delta V = IR$. The power dissipated is related to potential difference, current, and resistance by $P = VI = I^2R = V^2/R$.

Solve: **(a)** The resistance between the hands is

$$R = \rho L/A = \rho L(\pi d/4) = 4\rho/(\pi d^2) = 4(5.0\ \Omega\cdot\text{m})(1.6\text{ m})/(\pi(0.10\text{ m})^2) = 1.0\times10^3\ \Omega$$

(b) The potential difference across a resistor of this size when there is a current of 100 mA is

$$\Delta V = IR = (0.10\text{ A})(1.02\times10^3\ \Omega) = 100\text{ V}$$

Assess: This result tells us that a potential difference of 100 V can be lethal. This should give you new respect for the potential difference at all of your electrical outlets.

Q23.1. Reason: No, this is not a complete circuit. A connection from the outer metal case of the bulb to the negative terminal of the battery would complete the circuit and light the bulb.

Assess: A complete circuit can be made in a few different ways with just one wire, one bulb, and one battery.

In fact, if your bulb is rated for 9 V you can even do it without the wire since both terminals of a 9 V battery are on the same end.

Q23.3. Reason: (a) $I_{out} = I_{in}$ because of conservation of charge. There are no branch points and the charge carriers must go somewhere (they don't build up).

(b) As we saw in the previous part, the current must be the same everywhere in the wires and resistors. Since $R = \Delta V / I$ and the current is the same for all, then we look at the graph to see what the potential drop across each resistor is: $\Delta V_3 > \Delta V_1 > \Delta V_2$. Therefore $R_3 > R_1 > R_2$.

Assess: Greater resistances have a greater voltage drop across them for a given current.

Q23.5. Reason: Since the resistors are connected in parallel across the battery they have the same potential difference (voltage drop) across them. Hence, we use this version of the power equation: $P = \frac{(\Delta V)^2}{R}$.

The resistor with the lower resistance will dissipate more power (since R is in the denominator); that is R_2.

Assess: You might wonder why we don't apply $P = I^2 R$ for the power instead of $P = (\Delta V)^2 / R$; that would seem to give the opposite answer. It is because I is not the same for the two resistors, so $P = I^2 R$ doesn't allow a comparison based on R since we don't know the currents. We used $P = (\Delta V)^2 / R$ because we know that ΔV is the same for both resistors.

Think about this result in terms of light bulbs. The conclusion is that the resistance of a 100 W light bulb is lower that the resistance of a 60 W light bulb (assuming 120 V in both cases). Measure the resistance of some light bulbs with an ohmmeter.

Q23.7. Reason: (a) Since both points a and c are at the same potential as the positive terminal, and both points b and d are at the same potential as the negative terminal, then $\Delta V_{ab} = \Delta V_{cd}$.

(b) $I_1 > I_2 = I_3$

Assess: Because the three identical resistors in series have a total resistance three times that in the left circuit, the current in the right circuit will be 1/3 the current in the left circuit, and so the voltage drop across each of the three resistors will be 1/3 the voltage drop across the lone resistor, but the total potential difference will be $3 \times (1/3)$ and $\Delta V_{ab} = \Delta V_{cd}$, just as we found from our earlier reasoning.

Q23.11. Reason: Charges have more pathways with the three resistors than they would with just the $200\,\Omega$ resistor, so the total resistance must be less than $200\,\Omega$. Adding resistors in parallel *always* decreases the total resistance for this reason.

Assess: The result would be the same if only one of the unknown resistors were in parallel instead of two. We can check the original result by using the formula for the equivalent resistance. The three resistors in parallel are equivalent to a resistor of resistance R_{eq}.

$$\frac{1}{R_{\text{eq}}} = \frac{1}{R} + \frac{1}{200\,\Omega} + \frac{1}{R} = \frac{2}{R} + \frac{1}{200\,\Omega} = \frac{400\,\Omega + R}{(200\,\Omega)R}$$

$$R_{\text{eq}} = \frac{(200\,\Omega)R}{400\,\Omega + R} = \frac{200\,\Omega}{1 + \left(\dfrac{400\,\Omega}{R}\right)}$$

Now take limits: as $R \to 0\,\Omega$, $R_{\text{eq}} \to 0\,\Omega$, and as $R \to \infty\,\Omega$, $R_{\text{eq}} \to 200\,\Omega$. So R_{eq} is between $0\,\Omega$ and $200\,\Omega$.

Q23.13. Reason: The brightness of the bulbs is determined by how much power is dissipated in each bulb, so we use the power equation. When the resistances of the bulbs we are comparing are the same we compare currents through the bulbs: $P = I^2 R$.

Bulb A is in parallel with the series combination of B and C. Since bulbs B and C are in series, they have the same current through them, so they will be equally bright. The resistance of the B-C combination is greater than the resistance of bulb A, so the current through bulb A is greater than the current through bulbs B and C. Therefore $I^2 R$ for bulb A is larger than for bulbs B and C: $P_A > P_B = P_C$. This is the ranking of the brightness.

Assess: It often helps novices to re-draw the diagram to see more easily which bulbs are in parallel and which are in series.

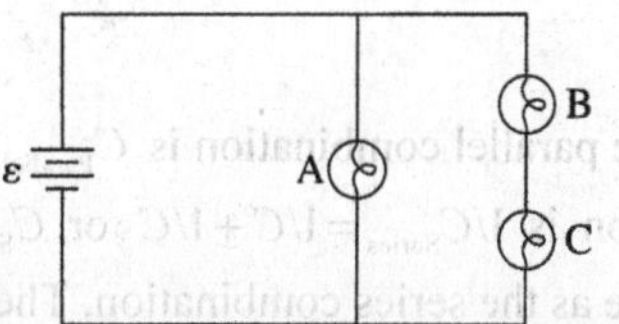

Q23.15. Reason: The brightness of the bulbs is determined by how much power is dissipated in each bulb, so we use the power equation. When the resistances of the bulbs we are comparing are the same (as is the case here) we compare currents through the bulbs.

$$P = I^2 R$$

Bulbs D and E have the same current through them so they are equally bright.

We need to examine the resistance of each major branch. They can be compared in one's head to arrive at $R_{\text{ABC}} < R_{\text{DE}}$ because of the parallel bulbs B and C. But we can also be more quantitative about it. Call the resistance of each bulb R; then the resistance of the ABC branch is $R_{\text{ABC}} = (3/2)R$ and the resistance of the DE branch is $R_{\text{DE}} = 2R$.

Because ΔV is the same across each major branch we can use Ohm's law over both branches $\Delta V = I_{\text{ABC}} R_{\text{ABC}} = I_{\text{DE}} R_{\text{DE}}$.

$$\frac{I_{\text{ABC}}}{I_{\text{DE}}} = \frac{R_{\text{DE}}}{R_{\text{ABC}}} \frac{2R}{(3/2)R} = \frac{4}{3}$$

The current in bulb A is I_{ABC} and so we see that bulb A is brighter than bulbs D and E.

$I_B = I_C = (1/2)I_A = (1/2)I_{\text{ABC}} = (2/3)I_{\text{DE}}$, so bulbs B and C are not as bright as bulbs D and E.

$P_A > P_D = P_E > P_B = P_C$. This is the ranking of the brightness.

Assess: When you have a thorough understanding and a bit of practice then this question can be answered almost at a glance. All of the details we worked through make sense intuitively.

Q23.21. Reason: (a) Because a good ammeter has very low resistance, the current would nearly all go through the ammeter, so the current in the $5.0\ \Omega$ resistor is nearly zero.

(b) Put the ammeter in series with the resistor rather than in parallel with it. The current will be the same anywhere in the one-loop circuit, so it doesn't matter exactly where the ammeter is placed in series.
Assess: Ammeters should be placed in series in the circuit where we want to know the current.

Q23.25. Reason: Given that the resistors are all identical, the pair in parallel (circuit A) has a lower resistance than the pair in series (circuit B). The current will be greater, therefore, in circuit A where the resistance is less. Since the current is greater in A, it will discharge the capacitor in the shortest amount of time.
Assess: Don't let the fact that the capacitor is drawn between the two resistors in circuit B fool you. The three elements are all still in series, and it would be the same if the resistors were drawn in series next to each other.

Q23.29. Reason: The resistance of an axon between one node and the next is fixed and fairly constant. The myelin insulation increases the separation between the inner conducting fluid and the outer conducting fluid. Since the capacitance of a capacitor depends inversely on the electrode spacing, the myelin reduces the capacitance of the membrane. Reducing the capacitance reduces the time constant. Since the time constant is a measure of how much time it takes for a signal to jump from one node to the next, an axon sheathed in myelin will cause faster signal propagation.
Assess: The true understanding of most biological systems requires knowledge of physics.

Q23.31. Reason: The power dissipated is determined by $P = I^2 R$. Since the resistors are in series, they each have the same current. As a result, the larger resistor will dissipate the most power. The correct choice is B.
Assess: If the current in two resistors is the same, the power dissipated will depend only on the resistance of the resistors.

Q23.35. Reason: The capacitance of the parallel combination is $C_{\text{Parallel}} = C + C = 2C$.

The capacitance of the series combination is $1/C_{\text{Series}} = 1/C + 1/C$ or $C_{\text{Series}} = C/2$. This calculation shows that the combination has four times the capacitance as the series combination. The correct choice is B.
Assess: This question requires a good knowledge of how capacitors add in parallel and in series.

Q23.37. Reason: Assume a spherical cell as in Examples 23.13 and 23.14. Reducing the diameter will decrease the surface area, which decreases the cross sectional area of the resistor (increasing the resistance) and decreases the area of the capacitor (decreasing the capacitance). Therefore the correct answer is B.
Assess: In $\tau = RC = \rho(L/A)(\varepsilon_0 A/d)$ the area cancels so τ remains unchanged.

Problems

P23.3. Prepare: Circuits are generally drawn with straight lines and 90° angles for the wires. We also usually orient the battery with the positive terminal up so the higher potential part of the circuit is toward the top of the paper.
Solve:

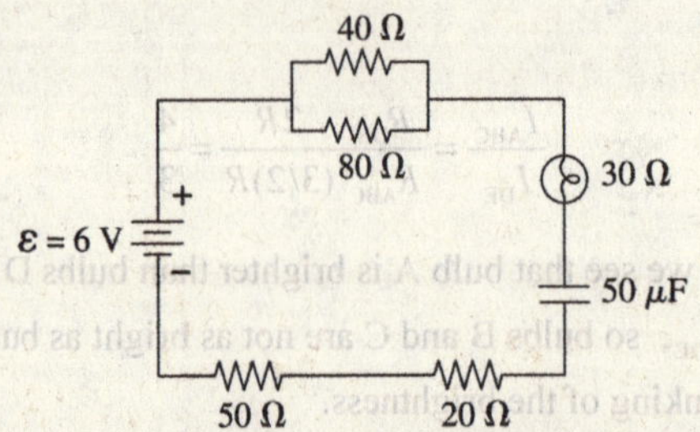

Assess: The elements that are in series with each other have the same current in them; the elements that are in parallel with each other have the same potential difference across them.

P23.5. Prepare: We will use Ohm's law ($\Delta V = IR$) for the resistor and the bulb, use Tactics Box 23.1 for Kirchoff's loop law, and assume that the connecting wires are ideal.
Solve: Let us assign clock direction to the current in the circuit.
(a): The Kirchoff's loop law is:

$$\Sigma(\Delta V)_i = \Delta V_{bat} + \Delta V_{resistor} + \Delta V_{bulb} = \mathcal{E} + (V_2 - V_1) + (V_3 - V_2) = 0$$
$$3.0\,\text{V} - I(2.0\,\Omega) - I(1.0\,\Omega) = 0 \Rightarrow I = 1.0\,\text{A}.$$

So,

$$\Delta V_{12} = V_2 - V_1 = -I(2.0\,\Omega) = -(1.0\,\text{A})(2.0\,\Omega) = -2.0\,\text{V}$$
$$\Delta V_{23} = V_3 - V_2 = -I(1.0\,\Omega) = -(1.0\,\text{A})(1.0\,\Omega) = -1.0\,\text{V}$$
$$\Delta V_{34} = V_4 - V_3 = 0\,\text{V}$$

The magnitudes are $\Delta V_{12} = 2.0\,\text{V}$, $\Delta V_{23} = 1.0\,\text{V}$, $\Delta V_{34} = 0\,\text{V}$.

(b) When the bulb is removed from the socket, no current flows in the circuit. Thus, $\Delta V_{12} = V_2 - V_1 = I(2.0\,\Omega) = (0\,\text{A})(2.0\,\Omega) = 0\,\text{V}$. This means that points 1 and 2 are at the same potential as the positive terminal of the battery. For the same reason, $\Delta V_{34} = V_4 - V_3 = 0$, implying that points 3 and 4 are at the same potential as the negative terminal of the battery. Finally, $\Delta V_{23} = V_3 - V_2 = V_4 - V_1 = -3.0\,\text{V}$. The magnitudes are $\Delta V_{12} = 0\,\text{V}$, $\Delta V_{23} = 3.0\,\text{V}$, $\Delta V_{34} = 0\,\text{V}$.
Assess: The potential at point 1 is higher than at point 4. This is what you would have expected.

P23.7. Prepare: Assume ideal connecting wires and an ideal battery for which $\Delta V_{bat} = E$. We assume a clockwise direction for I. Note that the choice of the current's direction is in general arbitrary; if we guess wrong the sign of the current will let us know. (In this simple case it is easy to see that the 6.0 V battery will triumph over the 3.0 V battery and the current will really go counter-clockwise.) Label the 3.0 V battery as 1 and the 6.0 V battery as 2.
Solve:
(a) Kirchhoff's loop law, going clockwise from the negative terminal of the 3.0 V battery is

$$\Sigma \Delta V_i = \Delta V_{bat\,1} + \Delta V_R + \Delta V_{bat\,2} = 0 = 3.0\,\text{V} - (18\,\Omega)I - 6.0\,\text{V}$$

$$I = \frac{3.0\,\text{V}}{-18\,\Omega} = -\frac{1}{6}\,\text{A} \approx -0.17\,\text{A}$$

The negative sign tells us the direction of the current is really to the left, contrary to our assumption.
(b)

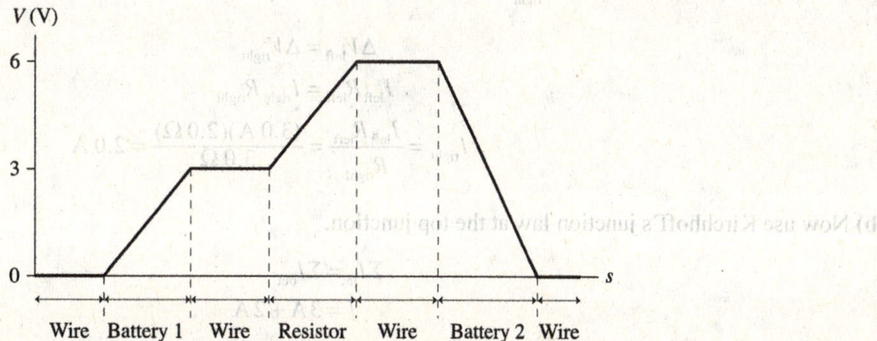

Assess: The final potential is the same as the initial potential, as required.

P23.11. Prepare: The three resistances in **(a)**, **(b)**, and **(c)** are parallel resistors. We will thus use Equation 23.12 to find the equivalent resistance.
Solve: (a) The equivalent resistance is

$$R_{eq} = \left(\frac{1}{2.0\,\Omega} + \frac{1}{3.0\,\Omega} + \frac{1}{6.0\,\Omega} \right)^{-1} = 1.0\,\Omega$$

(b) The equivalent resistance is

$$R_{eq} = \left(\frac{1}{3.0\,\Omega} + \frac{1}{3.0\,\Omega} + \frac{1}{3.0\,\Omega} \right)^{-1} = 1.0\,\Omega$$

(c) The equivalent resistance is

$$R_{eq} = \left(\frac{1}{2.0\,\Omega} + \frac{1}{1.0\,\Omega} + \frac{1}{2.0\,\Omega} \right)^{-1} = 0.50\,\Omega$$

Assess: We must learn how to combine series and parallel resistors.

P23.13. Prepare: Use Equation 23.12 to find the unknown resistance.
Solve:

$$\frac{1}{30\,\Omega} = \frac{1}{R} + \frac{1}{R} + \frac{1}{R} = \frac{3}{R} \Rightarrow R = 90\,\Omega$$

Assess: When the three $90\,\Omega$ resistors are added in series the equivalent resistance is $270\,\Omega$.

P23.15. Prepare: When resistors are combined in parallel, the combination is always less than the smallest resistor. As a result, the way to obtain the smallest resistance is to connect the resistors in parallel.
Solve: The equivalent resistance when the six resistors are connected in parallel is obtained by

$$\frac{1}{R_{Total}} = \frac{1}{R} + \frac{1}{R} + \frac{1}{R} + \frac{1}{R} + \frac{1}{R} + \frac{1}{R} = \frac{6}{R} \quad \text{or} \quad R_{Total} = R/6 = 1000\,\Omega/6 = 170\,\Omega$$

Assess: Notice that the total resistance is less than the smallest resistor. The smallest resistor is $1000\,\Omega$ and the total resistance when the six resistors are connected in parallel is $170\,\Omega$.

P23.21. Prepare: The two resistors are in parallel and so have the same voltage drop $\Delta V = IR$ across them: $\Delta V_{left} = \Delta V_{right}$.
Solve: (a) We want to solve for I_{right}.

$$\Delta V_{left} = \Delta V_{right}$$

$$I_{left} R_{left} = I_{right} R_{right}$$

$$I_{right} = \frac{I_{left} R_{left}}{R_{right}} = \frac{(3.0\,\text{A})(2.0\,\Omega)}{3.0\,\Omega} = 2.0\,\text{A}$$

(b) Now use Kirchhoff's junction law at the top junction.

$$\Sigma I_{in} = \Sigma I_{out}$$

$$I = 3\,\text{A} + 2\,\text{A}$$

So $I = 5.0\,\text{A}$.

Assess: We see that the resistor with the greater resistance has less current, as we would expect in this parallel situation.

When the two currents rejoin at the bottom junction the sum is again 5.0 A.

P23.23. Prepare: Assume that the connecting wire and the battery are ideal. The middle and right branches are in parallel, so the potential difference across these two branches must be the same.
Solve: The currents in the middle and right branches are known, so the potential differences across the two branches, using Ohm's law, are $\Delta V_{\text{middle}} = (3.0\text{ A})R = \Delta V_{\text{right}} = (2.0\text{ A})(R + 10\,\Omega)$.
This is easily solved to give $R = 20\,\Omega$. The middle resistor R is connected directly across the battery, thus (for an ideal battery, with no internal resistance) the potential difference ΔV_{middle} equals the emf of the battery. That is
$\mathcal{E} = \Delta V_{\text{middle}} = (3.0\text{ A})(20\,\Omega) = 60\text{ V}$.
Assess: Note that the potential difference across parallel resistors is the same.

P23.25. Prepare: Assume ideal batteries and wires and ohmic resistors. Label the top resistor as 1 and the other three from left to right 2, 3, and 4. The total resistance is

$$R_{\text{tot}} = 5.0\,\Omega + \frac{5.0\,\Omega}{3} = 6.667\,\Omega$$

Solve: The current through the battery (and R_1) is then $I = \Delta V / R_{\text{tot}} = (10\text{ V})/(6.667\,\Omega) = 1.5\text{ A}$. From the junction law and symmetry that current splits up evenly in the other three resistors, so $I_2 = I_3 = I_4 = 0.50\text{ A}$.
Now $\Delta V_1 = I_1 R_1 = (1.5\text{ A})(5.0\,\Omega) = 7.5\text{ V}$. The loop law using the battery, R_1, and R_2 shows that $\Delta V_2 = 10\text{ V} - 7.5\text{ V} = 2.5\text{ V}$ and similarly for ΔV_3 and ΔV_4.

R	I (A)	ΔV (V)
R_1	1.5	7.5
R_2	0.50	2.5
R_3	0.50	2.5
R_4	0.50	2.5

Assess: From symmetry we expect the results to be the same for R_2, R_3, and R_4.

P23.27. Prepare: The battery and the connecting wires are ideal. The figure shows how to simplify the circuit in the figure using the laws of series and parallel resistances. We have labeled the resistors as $R_1 = 6.0\,\Omega$, $R_2 = 15\,\Omega$, $R_3 = 6.0\,\Omega$, and $R_4 = 4.0\,\Omega$. Having reduced the circuit to a single equivalent resistance R_{eq}, we will reverse the procedure and "build up" the circuit using the loop law and the junction law to find the current and potential difference of each resistor.

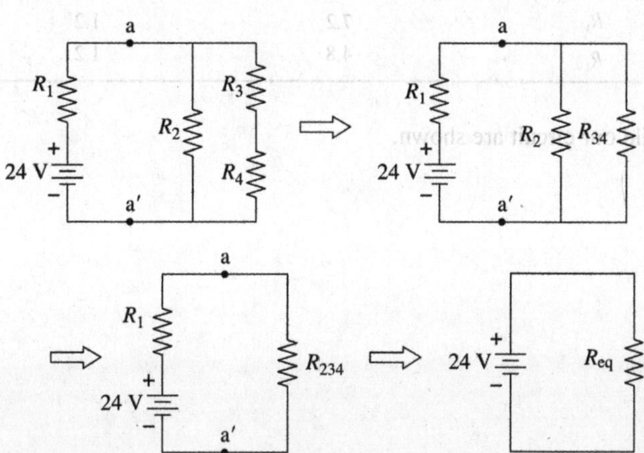

Solve: R_3 and R_4 are combined to get $R_{34} = 10\,\Omega$, and then R_{34} and R_2 are combined to obtain R_{234}:

$$\frac{1}{R_{234}} = \frac{1}{R_2} + \frac{1}{R_{34}} = \frac{1}{15\,\Omega} + \frac{1}{10\,\Omega} \Rightarrow R_{234} = 6\,\Omega$$

Next, R_{234} and R_1 are combined to obtain

$$R_{eq} = R_{234} + R_1 = 6.0\,\Omega + 6.0\,\Omega = 12\,\Omega$$

From the final circuit,

$$I = \frac{\mathcal{E}}{R_{eq}} = \frac{24\text{ V}}{12\,\Omega} = 2.0\text{ A}$$

Thus, the current through the battery and R_1 is $I_{R1} = 2.0$ A and the potential difference across R_1 is $I(R_1) = (2.0\text{ A})(6.0\,\Omega) = 12$ V.

As we rebuild the circuit, we note that series resistors *must* have the same current I and that parallel resistors *must* have the same potential difference ΔV.

In Step 1 of the previous figure, $R_{eq} = 12\,\Omega$ is returned to $R_1 = 6.0\,\Omega$ and $R_{234} = 6.0\,\Omega$ in series. Both resistors must have the same 2.0 A current as R_{eq}. We then use Ohm's law to find

$$\Delta V_{R1} = (2.0\text{ A})(6.0\,\Omega) = 12\text{ V} \qquad \Delta V_{R234} = (2.0\text{ A})(6.0\,\Omega) = 12\text{ V}$$

As a check, $12\text{ V} + 12\text{ V} = 24$ V, which was ΔV of the R_{eq} resistor. In Step 2, the resistance R_{234} is returned to R_2 and R_{34} in parallel. Both resistors must have the same $\Delta V = 12$ V as the resistor R_{234}. Then from Ohm's law,

$$I_{R2} = \frac{12\text{ V}}{15\,\Omega} = 0.8\text{ A} \qquad I_{R34} = \frac{12\text{ V}}{10\,\Omega} = 1.2\text{ A}$$

As a check, $I_{R2} + I_{R34} = 2.0$ A, which was the current I of the R_{234} resistor. In Step 3, R_{34} is returned to R_3 and R_4 in series. Both resistors must have the same 1.2 A as the R_{34} resistor. We then use Ohm's law to find

$$(\Delta V)_{R3} = (1.2\text{ A})(6.0\,\Omega) = 7.2\text{ V} \qquad (\Delta V)_{R4} = (1.2\text{ A})(4.0\,\Omega) = 4.8\text{ V}$$

As a check, $7.2\text{ V} + 4.8\text{ V} = 12$ V, which was ΔV of the resistor R_{34}.

Resistor	Potential difference (V)	Current (A)
R_1	12	2.0
R_2	12	0.8
R_3	7.2	1.2
R_4	4.8	1.2

The three steps as we rebuild our circuit are shown.

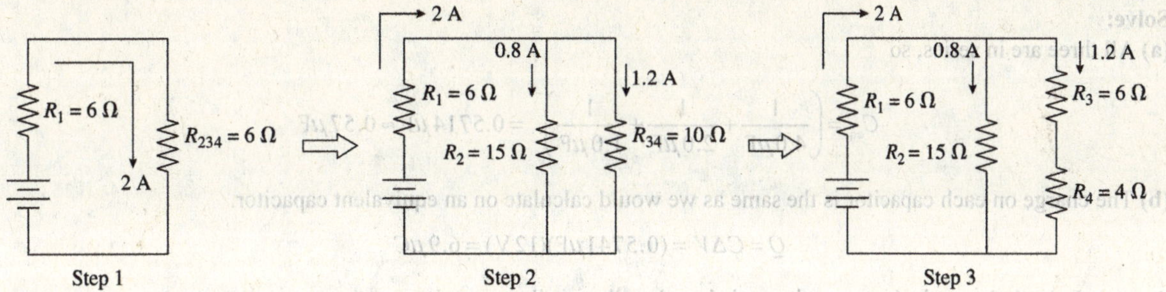

Assess: This problem requires a good understanding of how to first reduce a circuit to a single equivalent resistance and then to build up a circuit.

P23.33. Prepare: Capacitors in series follow Equation 23.19.
Solve: The equivalent capacitance is

$$C_{eq} = \left(\frac{1}{C_1} + \frac{1}{C_2} + \frac{1}{C_3}\right)^{-1} = \left(\frac{1}{6.0\ \mu F} + \frac{1}{10\ \mu F} + \frac{1}{16\ \mu F}\right)^{-1} = 3.04 \times 10^{-6}\ F = 3.0\ \mu F$$

Assess: This value is the smallest of all capacitances, so the result is reasonable.

P23.35. Prepare: Two capacitors in series combine to give less capacitance according to Equation 23.19.
Solve: Since we have a 75 μF capacitor and we want a 50 μF capacitance, we must connect the second capacitor in *series* with the 75 μF capacitor. The capacitance of the second capacitor is calculated as follows:

$$\frac{1}{75\ \mu F} + \frac{1}{C} = \frac{1}{50\ \mu F} \Rightarrow C = 150\ \mu F$$

Assess: We must learn how to combine series and parallel capacitances.

P23.37. Prepare: The following visual overview shows how to find the equivalent capacitance of the three capacitors shown in the figure.

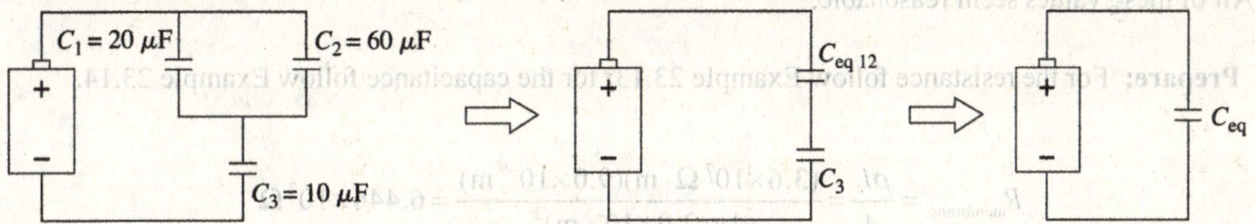

Solve: Because C_1 and C_2 are in parallel, their equivalent capacitance $C_{eq\,12}$ is

$$C_{eq\ 12} = C_1 + C_2 = 20\ \mu F + 60\ \mu F = 80\ \mu F$$

Then, $C_{eq\,12}$ and C_3 are in series. So,

$$\frac{1}{C_{eq}} = \frac{1}{C_{eq\ 12}} + \frac{1}{C_3} = \frac{1}{80\ \mu F} + \frac{1}{10\ \mu F} = \frac{9}{80}(\mu F)^{-1} \Rightarrow C_{eq} = \frac{80}{9}\ \mu F = 8.9\ \mu F$$

Assess: Learn well how to combine series and parallel capacitances.

P23.39. Prepare: For capacitors in series we know the equivalent capacitance is less than any of the individual capacitances. The charge on capacitors in series is the same.

Solve:
(a) All three are in series, so

$$C_{eq} = \left(\frac{1}{4.0\,\mu F} + \frac{1}{2.0\,\mu F} + \frac{1}{1.0\,\mu F} \right)^{-1} = 0.5714\,\mu F \approx 0.57\,\mu F$$

(b) The charge on each capacitor is the same as we would calculate on an equivalent capacitor.

$$Q = C\Delta V = (0.5741\,\mu F)(12\,V) = 6.9\,\mu C$$

Assess: Indeed, the equivalent capacitance is less than the smallest capacitor.

P23.45. Prepare: The capacitor discharges through a resistor. The switch in the circuit in P23.45 is in position a. When the switch is in position b the circuit consists of a capacitor and a resistor. Current and voltage during a capacitor discharge are given by Equations 23.22. Because the charge on a capacitor is $Q = C\Delta V$, the decay of the capacitor charge is given by $Q = Q_0\,e^{-t/\tau}$.

Solve: **(a)** The switch has been in position a for a long time. That means the capacitor is fully charged to a charge $Q_0 = C\Delta V = C\varepsilon = (2\,\mu F)(9\,V) = 18\,\mu C$.

Immediately after the switch is moved to the b position, the charge on the capacitor is $Q_0 = 18\,\mu C$. The current through the resistor is

$$I_0 = \frac{\Delta V_R}{R} = \frac{9\,V}{50\,\Omega} = 0.18\,A = 180\,mA$$

Note that as soon as the switch is closed, the potential difference across the capacitor ΔV_C appears across the $50\,\Omega$ resistor.

(b) The charge Q_0 decays as $Q = Q_0\,e^{-t/\tau}$, where $\tau = RC = (50\,\Omega)(2\,\mu F) = 100\,\mu s$.

Thus, the charge is $Q = (18\,\mu C)e^{-50\,\mu s/(100\,\mu s)} = (18\,\mu C)\,e^{-0.5} = 10.9\,\mu C = 11\,\mu C$.

The resistor current is $I = I_0\,e^{-t/\tau} = (180\,mA)e^{-50\,\mu s/(100\,\mu s)} = 110\,mA$.

(c) Likewise, the charge is $Q = 2.4\,\mu C$ and the current is $I = 24\,mA$.

Assess: All of these values seem reasonable.

P23.47. Prepare: For the resistance follow Example 23.13; for the capacitance follow Example 23.14.
Solve:

$$R_{membrane} = \frac{\rho L}{A} = \frac{(3.6\times10^7\,\Omega\cdot m)(9.0\times10^{-9}\,m)}{4\pi(2.0\times10^{-5}\,m)^2} = 6.446\times10^7\,\Omega$$

$$C_{membrane} = \frac{\kappa\varepsilon_0 A}{d} = \frac{9.0(8.85\times10^{-12}\,C^2/(N\cdot m^2))4\pi(2.0\times10^{-5}\,m)^2}{9.0\times10^{-9}\,m} = 4.448\times10^{-11}\,F$$

$$\tau = RC = (6.446\times10^7\,\Omega)(4.448\times10^{-11}\,F) = 2.9\,ms$$

Assess: Our result is similar to the one obtained in the chapter.

P23.49. Prepare: Follow the text and assume the conduction speed is approximately the distance between the nodes divided by the RC time constant.
Solve:

$$v = \frac{L_{node}}{\tau} = \frac{0.80\times10^{-3}\,m}{(20\,M\Omega)(1.2\,pF)} = 33\,m/s$$

Assess: This result is similar to the one in the chapter. The units do work out since $\Omega\cdot F = s$.

P23.53. Prepare: We will assume ideal connecting wires and an ideal power supply. The two light bulbs are basically two resistors in series. Our strategy is to find the current that flows through the two bulbs and then use Equation 22.13 to find the power dissipated by each bulb.

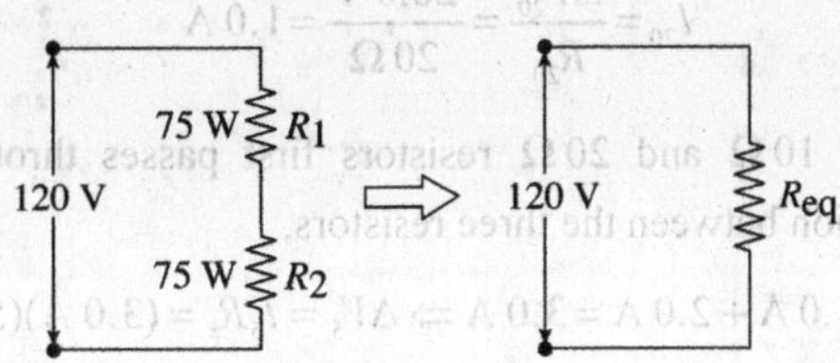

Solve: A 75 W (120 V) light bulb has a resistance of

$$R = \frac{\Delta V^2}{P} = \frac{(120 \text{ V})^2}{75 \text{ W}} = 192 \, \Omega$$

The combined resistance of the two bulbs is $R_{eq} = R_1 + R_2 = 192 \, \Omega + 192 \, \Omega = 384 \, \Omega$.

The current I flowing through R_{eq} is

$$I = \frac{\Delta V}{R_{eq}} = \frac{120 \text{ V}}{384 \, \Omega} = 0.3125 \text{ A}$$

Because R_{eq} is a series combination of R_1 and R_2, the current 0.3125 A flows through R_1 and R_2. Thus,

$$P_{R1} = I^2 R_1 = (0.3125 \text{ A})^2 (192 \, \Omega) = 19 \text{ W} = P_{R2}.$$

Assess: Note that a light bulb rated at 75 W (120 V) delivers 75 W power only when it is connected across a source of 120 V. As we see in this problem, the power delivered by two series bulbs across 120 V is considerably reduced compared to their rated power.

P23.55. Prepare: The $10 \, \Omega$ internal resistance of the battery is in series with the external $2.0 \, \Omega$ resistor. So the total resistance of the circuit is $1.0 \, \Omega + 2.0 \, \Omega = 3.0 \, \Omega$.

As a preliminary calculation, use Ohm's law to find the current in the circuit.

$$I = \frac{\Delta V}{R} = \frac{1.5 \text{ V}}{3.0 \, \Omega} = 0.50 \text{ A}$$

Solve: (a) The potential difference between the terminals of the battery is

$$\Delta V_{batt} = \mathcal{E} - I R_{internal} = 1.5 \text{ V} - (0.50 \text{ A})(1.0 \, \Omega) = 1.0 \text{ V}$$

(b) The total power dissipated is $P = I^2 R = (0.50 \text{ A})^2 (3.0 \, \Omega) = 0.75 \text{ W}$. The power dissipated internally in the battery is $P = I^2 R = (0.50 \text{ A})^2 (1.0 \, \Omega) = 0.25 \text{ W}$, or 1/3 of the total.

Assess: While ideal batteries are sources of constant emf, real batteries with internal resistance have a potential difference between the terminals that depends on the external resistance (the load).

P23.59. Prepare: The connecting wires are ideal, but the battery is not. We will designate the current in the $5.0 \, \Omega$ resistor I_5 and the voltage drop ΔV_5. Similar designations will be used for the other resistors.

Solve: Since the $10 \, \Omega$ resistor is dissipating 40 W,

$$P_{10} = I_{10}^2 R_{10} = 40 \text{ W} \Rightarrow I_{10} = \sqrt{\frac{P_{10}}{R_{10}}} = \sqrt{\frac{40 \text{ W}}{10 \, \Omega}} = 2.0 \text{ A} \Rightarrow \Delta V_{10} = I_{10} R_{10} = (2.0 \text{ A})(10 \, \Omega) = 20.0 \text{ V}$$

The $20\,\Omega$ resistor is in parallel with the $10\,\Omega$ resistor, so they have the same potential difference: $\Delta V_{20} = \Delta V_{10} = 20.0\,\text{V}$. From Ohm's law,

$$I_{20} = \frac{\Delta V_{20}}{R_{20}} = \frac{20.0\,\text{V}}{20\,\Omega} = 1.0\,\text{A}$$

The combined current through the $10\,\Omega$ and $20\,\Omega$ resistors first passes through the $5.0\,\Omega$ resistor. Applying Kirchhoff's junction law at the junction between the three resistors,

$$I_5 = I_{10} + I_{20} = 1.0\,\text{A} + 2.0\,\text{A} = 3.0\,\text{A} \Rightarrow \Delta V_5 = I_5 R_5 = (3.0\,\text{A})(5.0\,\Omega) = 15\,\text{V}$$

Knowing the currents and potential differences, we can now find the power dissipated:

$$P_5 = I_5 \Delta V_5 = (3.0\,\text{A})(15.0\,\text{V}) = 45\,\text{W} \qquad P_{20} = I_{20}\Delta V_{20} = (1.0\,\text{A})(20.0\,\text{V}) = 20\,\text{W}$$

P23.61. Prepare: The connecting wires are ideal with zero resistance. A visual overview of how to reduce the circuit to an equivalent resistance is shown.

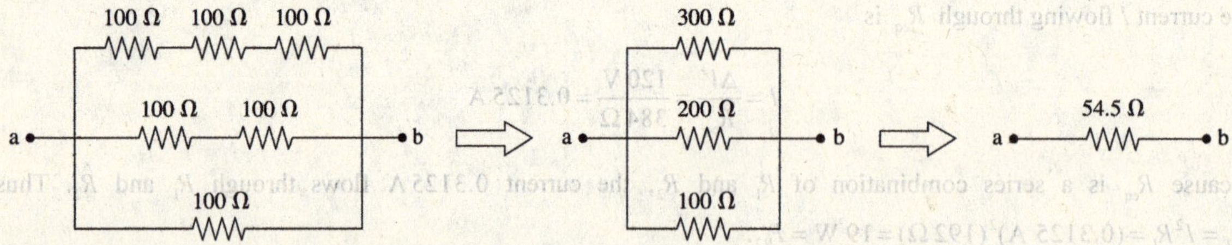

Solve: In the first step, the resistors $100\,\Omega$, $100\,\Omega$, and $100\,\Omega$ in the top branch are in series. Their combined resistance is $300\,\Omega$. In the middle branch, the two resistors, each $100\,\Omega$, are in series. So, their equivalent resistance is $200\,\Omega$. In the second step, the three resistors are in parallel. Their equivalent resistance is

$$\frac{1}{R_{eq}} = \frac{1}{300\,\Omega} + \frac{1}{200\,\Omega} + \frac{1}{100\,\Omega} \Rightarrow R_{eq} = 54.5\,\Omega$$

The equivalent resistance of the circuit is $55\,\Omega$.

P23.63. Prepare: Assume the battery is ideal. When two identical resistors are in series the equivalent resistance is $2R$, and when they are in parallel the equivalent resistance is $\frac{1}{2}R$.
Solve: Use $P = (\Delta V)^2 / R$ where the potential difference is the emf ε.

$$\frac{P_{parallel}}{P_{series}} = \frac{\varepsilon^2/(R_{eq})_{parallel}}{\varepsilon^2/(R_{eq})_{series}} = \frac{(R_{eq})_{series}}{(R_{eq})_{parallel}} = \frac{2R}{\frac{1}{2}R} = 4$$

Assess: When the resistance is lower the power dissipated is greater.

P23.67. Prepare: We will assume ideal connecting wires. Because the ammeter we have shows a full-scale deflection with a current of $500\,\mu\text{A}$, we must not pass a current greater than this through the ammeter.
Solve: The maximum potential difference is $5\,\text{V}$ and the maximum current is $500\,\mu\text{A}$. Using Ohm's law,

$$\Delta V = I_A R \Rightarrow 5.0\,\text{V} = (500\times10^{-6}\,\text{A})R \Rightarrow R = 10\,\text{k}\Omega.$$

P23.69. Prepare: Capacitors in parallel add to a greater capacitance compared to individual capacitances. On the other hand, capacitors in series add to a smaller capacitance compared to individual capacitances.

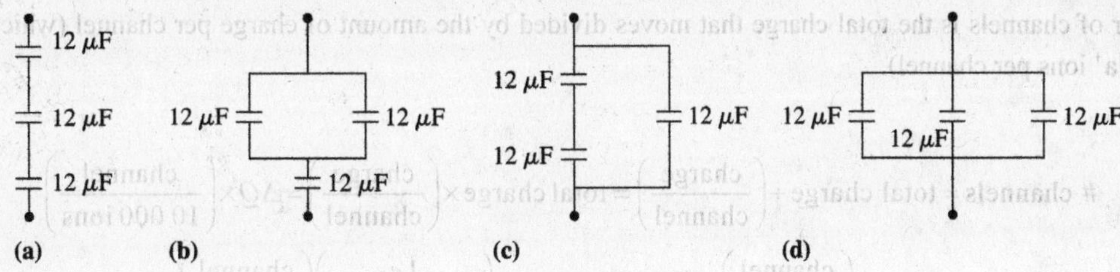

Solve: (a) Three capacitors in series:

$$\frac{1}{C_{eq}} = \frac{1}{12\ \mu F} + \frac{1}{12\ \mu F} + \frac{1}{12\ \mu F} = \frac{3}{12}(\mu F)^{-1} \Rightarrow C_{eq} = 4.0\ \mu F$$

(b) Two capacitors in parallel and the third in series with this parallel combination:

$$C_{eq\ 12} = 12\ \mu F + 12\ \mu F = 24\ \mu F \Rightarrow \frac{1}{C_{eq}} = \frac{1}{C_{eq\ 12}} + \frac{1}{12\ \mu F} = \frac{1}{24\ \mu F} + \frac{1}{12\ \mu F} = \frac{1}{8.0\ \mu F} \Rightarrow C_{eq} = 8.0\ \mu F$$

(c) Two capacitors in series and the third in parallel with this series combination:

$$\frac{1}{C_{eq\ 12}} = \frac{1}{12\ \mu F} + \frac{1}{12\ \mu F} = \frac{1}{6.0}\ (\mu F)^{-1} \Rightarrow C_{eq\ 12} = 6.0\ \mu F \Rightarrow C_{eq} = C_{eq\ 12} + 12\ \mu F = 6.0\ \mu F + 12\ \mu F = 18\ \mu F$$

(d) Three capacitors in parallel:

$$C_{eq} = 12\ \mu F + 12\ \mu F + 12\ \mu F = 36\ \mu F$$

Assess: Learn how to combine series and parallel capacitances.

P23.73. Prepare: A capacitor discharges through a resistor and the wires are ideal. Current and voltage during a capacitor discharge are given by Equations 23.22. Because the charge on a capacitor is $Q = C\Delta V$, the decay of the capacitor charge is given $Q = Q_0 e^{-t/\tau}$.

Solve: A capacitor initially charged to Q_0 decays as $Q = Q_0 e^{-t/\tau}$. We wish to find R so that a $1.0\ \mu F$ capacitor will discharge to 10% of its initial value in 2.0 ms. That is,

$$(0.10)Q_0 = Q_0\, e^{-2\ ms/(R(1\ \mu F))} \Rightarrow \ln(0.10) = -\frac{2\times10^{-3}\ s}{R(1.0\times10^{-6}\ F)} \Rightarrow R = \frac{-2.0\times10^{-3}\ s}{(1.0\times10^{-6}\ F)\ln(0.10)} = 870\ \Omega$$

Assess: A time constant of $\tau = RC = (870\ \Omega)(1.0\times10^{-6}\ F) = 0.87\ ms$ is reasonable.

P23.75. Prepare: A capacitor discharges through a resistor and we will assume that the wires are ideal. The discharge current or the resistor current follows Equation 23.22: $I = I_0 e^{-t/(RC)}$.

Solve: We wish to find the capacitance C so that the resistor current will decrease to 25% of its initial value in 2.5 ms. That is,

$$0.25\ I_0 = I_0\, e^{-2.5\ ms/((100\ \Omega)C)} \Rightarrow \ln(0.25) = -\frac{2.5\times10^{-3}\ s}{(100\ \Omega)C} \Rightarrow C = 18\ \mu F$$

Assess: A capacitance of $18\ \mu F$ is typical of capacitors.

P23.79. Prepare: The chapter says that when the sodium channels open, the potential inside the cell changes from $-70\ mV$ to $+40\ mV$ relative to outside the cell. This $\Delta V = 110\ mV$ will be accompanied by a movement of charge $\Delta Q = C\Delta V$ as the ions flow into the cell.

The number of channels is the total charge that moves divided by the amount of charge per channel (which is given as $10\,000$ Na^+ ions per channel).

Solve:

$$\text{\# channels} = \text{total charge} \div \left(\frac{\text{charge}}{\text{channel}}\right) = \text{total charge} \times \left(\frac{\text{charge}}{\text{channel}}\right) = \Delta Q \times \left(\frac{\text{channel}}{10\,000\,\text{ions}}\right)$$

$$= C\Delta V \times \left(\frac{\text{channel}}{10\,000\,e}\right) = (89\,\text{pF})(110\,\text{mV})\left(\frac{1\,e}{1.6\times10^{-19}\,\text{C}}\right)\left(\frac{\text{channel}}{10\,000\,e}\right)$$

$$= (9.79\times10^{-12}\,\text{C})\left(\frac{1\,e}{1.6\times10^{-19}\,\text{C}}\right)\left(\frac{\text{channel}}{10\,000\,e}\right) = 6100\,\text{channels}$$

Assess: Example 23.15 says there are "a great many channels" in a cell, and our answer bears this out.

P23.81. Prepare: The capacitance of a dielectric-filled capacitor is $C = \kappa\mathcal{E}_0 A/d$ and then Equation 21.17 gives the charge: $Q = C\Delta V_\text{C}$.

In Example 23.14 we are told the dielectric is $\kappa = 9.0$. We are also given $A = 6.0\times10^{-9}\,\text{m}^2$, $d = 7.0\times10^{-9}\,\text{m}$, and $\Delta V_\text{C} = 70\times10^{-3}\,\text{V}$.

The sodium ions have a $+1$ charge.

Solve: (a) We can combine both equations.

$$Q = C\Delta V_\text{C} = \frac{\kappa\mathcal{E}_0 A}{d}\Delta V_\text{C} = \frac{(9)(8.85\times10^{-12}\,\text{C}^2/\text{N}\cdot\text{m}^2)(6.0\times10^{-9}\,\text{m}^2)}{7.0\times10^{-9}\,\text{m}}(70\times10^{-3}\,\text{V}) = 4.78\times10^{-12}\,\text{C}$$

which should be reported as 4.8 pC to two significant figures.

(b)

$$4.78\times10^{-12}\,\text{C}\left(\frac{1\,e}{1.6\times10^{-19}\,\text{C}}\right) = 3.0\times10^{7}\,e$$

or 3.0×10^{7} sodium ions.

Assess: This tiny fraction of a Coulomb corresponds to 30 million sodium ions. There is a lot going on in each living cell!

24

MAGNETIC FIELDS AND FORCES

Q24.3. Reason: A compass points near the geographic north because the earth's magnetic field lines run from near the geographic South Pole (magnetic north pole) to near the geographic North Pole (magnetic south pole): see Figure 24.8 in textbook. The field is not dependent on the hemisphere one is standing in.
Assess: In a bar magnet magnetic field lines run from a north pole to a south pole. To find the geographic North Pole one uses the north pole of a magnet because the north pole will point in the direction of the earth's magnetic south pole.

Q24.5. Reason: Magnetotactic bacteria use the earth's field to navigate up and down in bodies of water. In the northern hemisphere, the vertical component of the field points down; bacteria seeking deep oxygen-poor water go with the field. In the southern hemisphere, the vertical component of the field points up, so bacteria following the field lines going with the field would go in the wrong direction, and likely not survive.
Assess: A number of species use magnetism for navigation; some of them migrate between hemispheres.

Q24.9. Reason: Conventional household wiring is composed of at least two current-carrying conductors. One conductor carries current to some appliance or light, and another conductor serves as the electrical return. However, household current reverses its direction about 120 times each second, or 60 cycles (reverse and back) each second.
Assess: It is unlikely you could detect current in the walls of a house with a compass for two reasons.
The first reason is if the direction of the current changes 120 times a second, the magnetic field surrounding the current-carrying wire changes direction 120 times each second. The mass inertia of the needle would prevent it from flipping 120 each second.
The second reason a compass would make a poor instrument to find wiring is that since household wiring is composed of a pair of current-carrying conductors close to each other, and the two currents will always be equal in magnitude, yet opposite in direction with each other, their magnetic fields will cancel.

Q24.11. Reason: Assuming the wire with the current directed to the right is the bottom wire, the bottom current-carrying wire produces a magnetic field pointing toward the bottom of the page at a point halfway between them. The top current-carrying wire produces a magnetic field pointing right at a point halfway between them. The total magnetic field is the resultant vector of the two magnetic fields as shown.

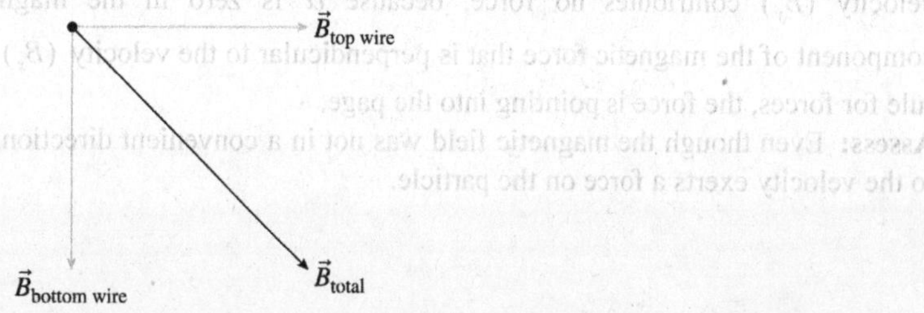

Assess: Whenever there are two magnetic fields acting at the same point, use vector addition to add the fields and find the resultant vector. In this case, the resultant vector is conceptually shown in the previous figure.

Q24.13. Reason: If a current is present in a long solenoid, the magnetic field that is created is uniform along its central axis. The direction of the magnetic field is in the direction the thumb on your right hand points when your fingers are curled in the direction of the current in accordance with the right-hand rule for fields. The magnetic field points to the right for all three compasses.

Assess: The right-hand rule can be used to determine the direction of the magnetic field due to any current, even a curved current.

Q24.17. Reason: Use the right-hand rule for forces to determine the direction of the magnetic field. See Figure 24.27. **(a)** The velocity points to the right and the force points down, so the magnetic field points out of the page. **(b)** The velocity points to the right and the force points into the page, so the magnetic field should point to the bottom of the page. BUT, because the charged particle is negative, the direction of the magnetic field is opposite. So the field points to the top of the page.

Assess: The direction of the force on a positive particle is governed by the right-hand rule. Conversely, the force on a negative particle is opposite.

Q24.19. Reason: The direction of the magnetic field of a current-carrying wire is described by the right-hand rule. Using this we see that above the wire the magnetic field is coming out of the page. See Figure 24.28. The force is found by the right-hand rule. For an electron, the force points up.

Assess: Magnetic force depends on charge, velocity, magnetic field, and direction. If you change the sign of the charge the direction of the force is opposite.

Q24.21. Reason: At the earth's magnetic north pole, the field lines come straight up out of the ground, as in the figure. From the right-hand rule, the force on a positive charge placed at point P and moving up the page would be to the right. So the force on an electron would be to the left. Thus the electron is steered to the left and moves counterclockwise, as viewed from above.

Assess: The earth's magnetic north pole is in the southern hemisphere so the revolution of an electron as described in this problem is characteristic of the aurora australis, or southern lights.

Q24.25. Reason: The way you find the direction of the magnetic field from a current carrying wire is shown in Section 24.3, using the right-hand rule for fields. The magnetic field for a current coming out of the page is curling around the wire in the counterclockwise direction.

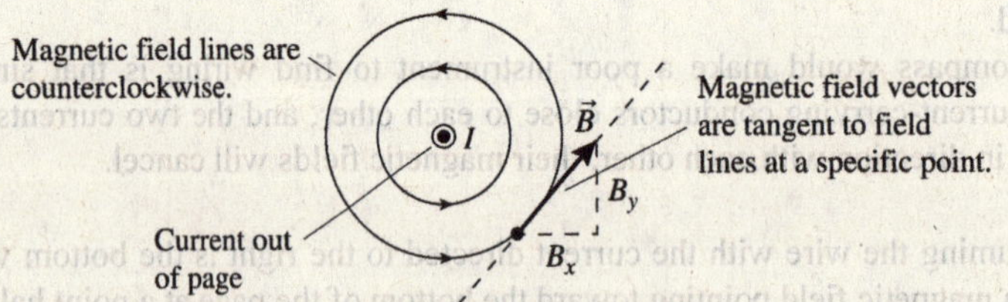

The magnetic field vector can be split into two components, B_x and B_y. The component that is parallel to the velocity (B_y) contributes no force, because α is zero in the magnetic force equation $F = qvB \sin \alpha$. The component of the magnetic force that is perpendicular to the velocity (B_x) contributes a force, and by the right-hand rule for forces, the force is pointing into the page.

Assess: Even though the magnetic field was not in a convenient direction, only the component that is perpendicular to the velocity exerts a force on the particle.

Q24.27. Reason: The force a magnetic field exerts on a current-carrying wire is given by Equation 24.9: $F_{wire} = ILB \sin \alpha$. In a solenoid the magnetic field inside runs parallel to the axis. If the wire runs down the middle of the solenoid then $\alpha = 0$ and $\sin 0 = 0$ so there is no force on the wire due to the field of the solenoid.

Assess: This situation could also be analyzed by looking at the moving charges in the wire and applying Equation 24.5.

Q24.31. Reason: Using the right-hand rule for fields, we conclude that the field at the point in question, due to the wire on the left is directed up the page. The field due to the wire on the right is also up the page at the midpoint. So the total field, which is the sum of these two, will also be up the page. The answer is A.

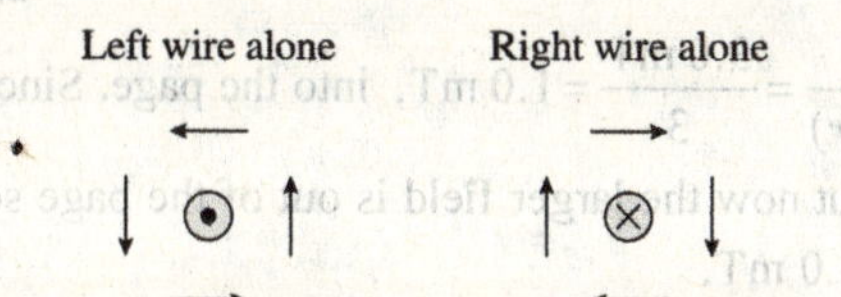

Assess: The orientation of the currents was very important here. For example, if both currents had been directed out of the page, then the field due to the wire on the left would still be up the page, but the field due to the wire on the right would be down the page. The total field would be zero.

Q24.33. Reason: When a charged particle enters a uniform magnetic field it moves in uniform circular motion. The radius of the curvature is determined by $r = mv/(qB)$, so the larger the radius, the larger the velocity. From the choices given, D has the largest radius, so it also has the highest velocity. The correct choice is D.

Problems

P24.1. Prepare: From $B = \dfrac{\mu_0 I}{2\pi r}$ we see that the field strength is inversely proportional to the distance from a long straight wire. We are told the field at 10 cm is $2\,\mu T$.

Solve: The field 1/10 the distance away would be 10 times as strong. So at 1 cm the field is $20\,\mu T$.

Assess: We do expect the field to be stronger near the current-carrying wire.

P24.3. Prepare: The equations for the magnetic field at the center of a loop and a wire are
$$B_{loop\ center} = \mu_0 I/(2R) \quad \text{and} \quad B_{wire} = \mu_0 I/(2\pi R)$$

Solve: (a) The radius of the loop is 0.5 cm and the magnetic field is 2.5 mT so the current is:
$$B_{loop\ center} = \frac{\mu_0 I}{2R} \Rightarrow I = \frac{2RB_{loop\ center}}{\mu_0} = \frac{2(0.5\times10^{-2}\ \text{m})(2.5\times10^{-3}\ \text{T})}{4\pi(10^{-7}\ \text{T}\cdot\text{m/A})} = 19.89\ \text{A}$$
which will be reported as 20 A.

(b) For a long, straight wire that carries a current I, the magnetic field strength is
$$B_{wire} = \frac{\mu_0 I}{2\pi d} \Rightarrow 2.50\times10^{-3}\ \text{T} = \frac{4\pi(10^{-7}\ \text{T}\cdot\text{m/A})(19.89\ \text{A})}{2\pi d} \Rightarrow d = 1.6\times10^{-3}\ \text{m}$$

Assess: The result for d obtained is consistent with the equations for B_{wire} and $B_{loop\ center}$.

P24.5. Prepare: To find the field at any point, we need to find the field we would have if only the top wire were present and the field we would have if only the bottom wire were present. Let the point where the field is given be called P and let the distance from point P to the bottom wire be called d. At P, the field due to the lower wire is, from the right-hand rule, out of the page. The field at P due to the top wire is into the page. Thus the field strength due to the top wire should be subtracted from the field strength due to the bottom wire (the bottom wire contributes a larger field since point P is closer to the bottom wire):
$$B_{total} = B_{bottom} - B_{top} = \frac{\mu_0 I}{2\pi r} - \frac{\mu_0 I}{2\pi(3r)} = \frac{\mu_0 I}{3\pi r}$$

This field is directed out of the page. We know that at point P, $B_{total} = 2.0$ mT $= \dfrac{\mu_0 I}{3\pi r}$, from which it follows that

$\dfrac{\mu_0 I}{2\pi r} = 3.0$ mT. Thus $B_{bottom} = \dfrac{\mu_0 I}{2\pi r} = 3.0$ mT.

Solve: At point 1, the field due to the bottom wire is into the page (from the right-hand rule) and has magnitude

$\dfrac{\mu_0 I}{2\pi r} = 3.0$ mT. Since the top and bottom wires have the same current and are equidistant to point 1, the field at point

1 due to the top wire is also 3.0 mT and from the right-hand rule, is also into the page. Thus the total field at point 1

is 6.0 mT, into the page. Finally, at point 2, the field due to the top wire is $\dfrac{\mu_0 I}{2\pi r} = 3.0$ mT, out of the page and the

field due to the bottom wire is $\dfrac{\mu_0 I}{2\pi(3r)} = \dfrac{3.0 \text{ mT}}{3} = 1.0$ mT, into the page. Since the fields are in opposite directions,

they are subtracted, as at point P, but now the larger field is out of the page so the total field is out of the page and

has magnitude: 3.0 mT $- 1.0$ mT $= 2.0$ mT.

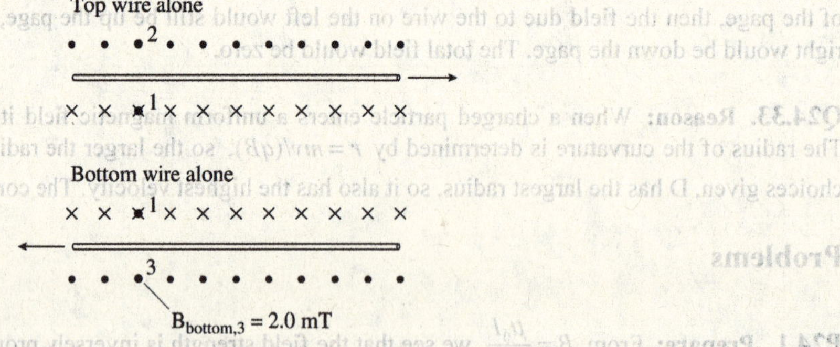

Assess: You might have guessed, using symmetry, that the field at point 2 would be the same as the field at point P. This is because these two points have in common that they are a distance r from one current, I, and a distance $3r$ from an equal current flowing in the opposite direction.

P24.7. Prepare: From $B = \dfrac{\mu_0 I}{2\pi r}$ for a long straight wire we see that the field is proportional to the current and

inversely proportional to the distance from the wire. Assume the distances given are from the center of the wire.
Solve: Increasing the distance by a factor of 5 would decrease the B field by the same factor, so we need to also increase the current by a factor of 5 to keep the field the same.
Assess: The answer seems reasonable.

P24.11. Prepare: Assume that the superconducting niobium wire is very long.
Solve: The magnetic field of a long wire carrying current I is

$$B_{wire} = \dfrac{\mu_0 I}{2\pi r}$$

We're interested in the magnetic field of the current right at the surface of the wire, where $r = 1.5$ mm $= 1.5 \times$

10^{-3} m. The maximum field is 0.10 T, so the maximum current is

$$I = \dfrac{(2\pi r)B_{wire}}{\mu_0} = \dfrac{2\pi(1.5 \times 10^{-3} \text{ m})(0.10 \text{ T})}{4\pi \times 10^{-7} \text{ T} \cdot \text{m/A}} = 750 \text{ A}$$

Assess: The current density in this superconducting wire is of the order of 1×10^8 A/m^2 (or current/area $=$

750 A/$(\pi(1.5 \times 10^{-3})^2$ m$^2) \cong 1.06 \times 10^8$ A/m^2). This is a typical value for conventional superconducting materials.

P24.13. Prepare: The magnetic field in the center of a solenoid is $B = \mu_0 IN/L$. This equation is independent of the radius of the solenoid, and therefore only depends on the current (I), number of loops (N), and the length of the solenoid (L).

Solve: Rewriting the equation for a solenoid, one can calculate the current in the wire to be the following:

$$B = \frac{\mu_0 IN}{L} \Rightarrow I = \frac{BL}{\mu_0 N}$$

$$I = \frac{(1.0 \text{ T})(2 \text{ m})}{(4\pi \times 10^{-7} \text{ T} \cdot \text{m/A})(1000)} = 1.6 \times 10^3 \text{ A} = 1.6 \text{ kA}$$

Assess: This current is very high. It is about 1000 times greater than a normal electrical current.

P24.15. Prepare: If we model the current as a single ($N = 1$) circular current loop we will want to use Equation 24.2.

$$B = \frac{\mu_0 I}{2R}$$

We are given $B = 3.0 \times 10^{-12}$ T and $R = 0.080$ m, we are asked to find I.

Solve: Solve for I.

$$I = \frac{2RB}{\mu_0} = \frac{2(0.080 \text{ m})(3.0 \times 10^{-12} \text{ T})}{4\pi \times 10^{-7} \text{ T} \cdot \text{m/A}} = 3.8 \times 10^{-7} \text{ A}$$

Assess: This is a small current, but it seems reasonable for a current in one's head. After all, the magnetic field produced is also very small.

Notice the units; both m and T cancel to leave A.

P24.21. Prepare: Assuming the electron is a single particle traveling in a circular orbit, the moving electron is, by definition, a circular current. Any current moving in a circle creates a magnetic field at the center of the orbit.

Solve: The current from a single electron moving in a circular orbit is

$$I = \frac{q}{t} = \frac{qv}{d} = \frac{qv}{2\pi r} = \frac{(1.60 \times 10^{-19} \text{ C})(2.2 \times 10^6 \text{ m/s})}{2\pi(5.3 \times 10^{-11} \text{ m})} = 1.06 \text{ mA}$$

With this current, the magnetic field can be calculated as

$$B = \frac{\mu_0 I}{2r} = \frac{(4\pi \times 10^{-7} \text{ T} \cdot \text{m/A})(1.06 \times 10^{-3} \text{ A})}{2(5.3 \times 10^{-11} \text{ m})} = 12.5 \text{ T}$$

which we'll report as 13 T.

Assess: This is a very large magnetic field strength, but note the extremely small radius of the electron orbit.

P24.23. Prepare: A magnetic field exerts a magnetic force on a moving charge. The magnitude of the magnetic force is calculated using the equation $F_B = |q|vB \sin \alpha$. The right-hand rule for forces is used to find the direction on a positive charge.

Solve: (a) The magnitude of the magnetic force on the charge is

$$F_B = |q|vB \sin \alpha = (1.60 \times 10^{-19} \text{ C})(1.0 \times 10^7 \text{ m/s})(0.50 \text{ T})(\sin 90°) = 8.0 \times 10^{-13} \text{ N}$$

Using the right-hand rule for forces, you find that the direction of the force is in the $+z$-direction or up. So, on the electron the force is in the $-z$-direction, or down.

(b) The magnitude of the magnetic force on the charge is

$$F_B = |q|vB \sin \alpha = (1.60 \times 10^{-19} \text{ C})(1.0 \times 10^7 \text{ m/s})(0.50 \text{ T})(\sin 90°) = 8.0 \times 10^{-13} \text{ N}$$

Using the right-hand rule for forces, you find that the direction of the force is in the $+yz$-direction. So, on the electron the force is in the $y - z$ plane, $45°$ clockwise from the $-z$-direction.

Assess: The velocity, magnetic field, and magnetic force are always perpendicular to each other. Using the right-hand rule for forces, you can determine the direction of the magnetic force without calculating all the individual components of velocity (v) and magnetic field (B).

P24.25. Prepare: Charged particles moving perpendicular to a uniform magnetic field undergo uniform circular motion ($a = v^2/r$) at a constant speed (v): $F_{net} = F_B \Rightarrow mv^2/r = qvB$ or $r = vm/(qB)$.

Solve: (a) The calculation for the radius of the circular orbit of the electron is

$$r_{electron} = \frac{vm}{qB} = \frac{(1.0 \times 10^6 \text{ m/s})(9.11 \times 10^{-31} \text{ kg})}{(1.60 \times 10^{-19} \text{ C})(5 \times 10^{-5} \text{ T})} = 1.14 \times 10^{-1} \text{ m} = 11 \text{ cm}$$

(b) And for the proton it is

$$r_{proton} = \frac{vm}{qB} = \frac{(5.0 \times 10^4 \text{ m/s})(1.67 \times 10^{-27} \text{ kg})}{(1.60 \times 10^{-19} \text{ C})(5 \times 10^{-5} \text{ T})} = 10 \text{ m}$$

Assess: In a magnetron, the charged particles have an angular velocity related to the linear velocity by $v = \omega r = 2\pi f r$, where f is the frequency and r is the radius of the circular motion. These charged particles create light that can be seen at the North Pole and South Pole, and the patterns are from these orbits.

P24.29. Prepare: Carefully examine Example 24.9 as a lead-in to this problem. Instead of the time (period) for one orbit we want the frequency, and $f = 1/T$.

$$f = \frac{|q|B}{2\pi m}$$

We are given $f = 2.4 \text{ GHz}$, and we look up the mass of the electron: $m = 9.11 \times 10^{-31}$ kg.

Solve: Solve the equation for B.

$$B = \frac{2\pi m f}{|q|} = \frac{2\pi (9.11 \times 10^{-31} \text{ kg})(2.4 \text{ GHz})}{1.6 \times 10^{-19} \text{ C}} = 0.086 \text{ T}$$

Assess: The emitted electromagnetic waves have a wavelength of 12.5 cm and are just right to excite the water molecules and cook the food.

Check the units, $1 \text{ C} = 1 \text{ A} \cdot \text{s}$ and $1 \text{ Hz} = 1 \text{ s}^{-1}$, so the units combine to give T.

P24.31. Prepare: The charge on a proton is $q = 1.6 \times 10^{-19}$ C and the mass is $m = 1.67 \times 10^{-27}$ kg.

Solve: (a) First solve for the speed from the kinetic energy.

$$K = \frac{1}{2} mv^2 \Rightarrow v = \sqrt{\frac{2K}{m}} = \sqrt{\frac{2(13 \text{ MeV})}{(1.67 \times 10^{-27} \text{ kg})} \frac{(1.6 \times 10^{-19} \text{ J})}{1 \text{ eV}}} = 4.99 \times 10^7 \text{ m/s}$$

Use Equation 24.8.

$$r = \frac{mv}{|q|B} = \frac{(1.67 \times 10^{-27} \text{ kg})(4.99 \times 10^7 \text{ m/s})}{(1.60 \times 10^{-19} \text{ C})(1.2 \text{ T})} = 43 \text{ cm}$$

(b) The frequency is the cycles per second.

$$f = \frac{|q|B}{2\pi m} = \frac{(1.60 \times 10^{-19} \text{ C})(1.2 \text{ T})}{2\pi (1.67 \times 10^{-27} \text{ kg})} = 18 \text{ MHz}$$

So it complete 18 million orbits in one second, and therefore complete 18,000 orbits in 1.0 ms.

Assess: The strong magnetic field makes the frequency very high.

P24.35. Prepare: To find the force on one of the 1.0 m long segments, we need the formula for the force on a current carrying wire in a magnetic field: $F = BIL$. The figure shows the field created by the wire on the left and the force felt by the wire on the right. We also need to know the magnetic field experienced by one of the wires (they experience the same field strength). This is given by $B = \dfrac{\mu_0 I}{2\pi r}$.

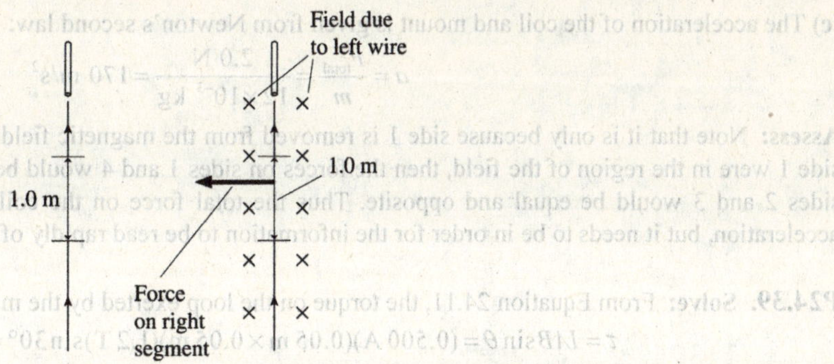

Solve: Combining the formula for the field experienced by a wire with the formula for the force experienced by the wire gives us the result we want:

$$F = BIL = \left(\frac{\mu_0 I}{2\pi r}\right) IL = \frac{\mu_0 I^2 L}{2\pi r} = \frac{(4\pi \times 10^{-7}\ \text{T} \cdot \text{m/A})(1.0\ \text{A})^2 (1.0\ \text{m})}{2\pi (1.0\ \text{m})} = 2.0 \times 10^{-7}\ \text{N}$$

Assess: This is actually how the amp is defined. You take two long wires and run identical current through them. Then if you gradually vary the current until the force per unit meter on the wires is 2×10^{-7} N/m, you know that 1 amp is flowing through the wires. All the units in electricity relate back to the amp. For example, the Coulomb is the amount of charge which flows past a point in a wire in one second if the current in the wire is one amp.

P24.37. Prepare: This problem makes us review some basic concepts in electricity. We will need to find the current in the coil using Ohm's law. Then to find the force on the coil, we will need to use the formula for the force on a current carrying wire: $F = BIL$. Finally, to find the acceleration of the coil, we will need Newton's second law.

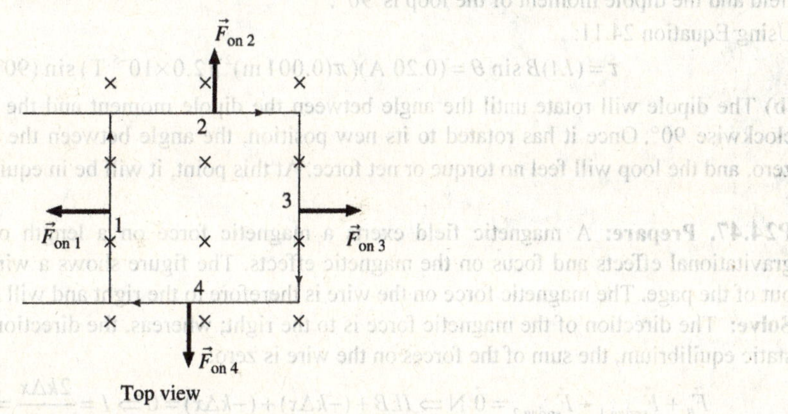

Solve: (a) From Ohm's law we can get the current in the coil:

$$I = \frac{\Delta V}{R} = \frac{5.0\ \text{V}}{1.5\ \Omega} = 3.33\ \text{A}$$

which rounds to $3.3\ \text{A}$.

(b) Notice in the figure that the four sides of the coil have been numbered 1 through 4. The force on the segments of wire on side 1 is zero because they are not in the magnetic field. The force on the segments on sides 2 and 3 is harder to find because these sides are only partially removed from the field. However, by the right-hand rule for forces, the force on the side 2 segments is up the page, while the force on the side 3 segments is down the page, so the sum of the force on side 2 and the force on side 3 is zero. Consequently, the total force on the coil is merely the force on side 4. The force on a single segment of side 4 is given by:

$$F = BIL = (0.30\ \text{T})(3.33\ \text{A})(1.0 \times 10^{-2}\ \text{m}) = 1.0 \times 10^{-2}\ \text{N}$$

Side 4 contains 200 such segments so the total force on the coil is $F_{\text{total}} = 200(1.0 \times 10^{-2}\ \text{N}) = 2.0\ \text{N}$.

(c) The acceleration of the coil and mount is given from Newton's second law:

$$a = \frac{F_{total}}{m} = \frac{2.0\ \text{N}}{12 \times 10^{-3}\ \text{kg}} = 170\ \text{m/s}^2$$

Assess: Note that it is only because side 1 is removed from the magnetic field that there is a net force on the coil. If side 1 were in the region of the field, then the forces on sides 1 and 4 would be equal and opposite and the forces on sides 2 and 3 would be equal and opposite. Thus the total force on the coil would be zero. This is a very large acceleration, but it needs to be in order for the information to be read rapidly off the hard drive.

P24.39. Solve: From Equation 24.11, the torque on the loop exerted by the magnetic field is

$$\tau = IAB\sin\theta = (0.500\ \text{A})(0.05\ \text{m} \times 0.05\ \text{m})(1.2\ \text{T})\sin 30° = 7.5 \times 10^{-4}\,\text{N} \cdot \text{m}.$$

P24.41. Prepare: The wire will create a magnetic field at the position of the loop, so there will be a torque on the loop. The magnetic field created by the wire at the position of the loop will be approximately uniform, since the distance between the wire and the loop is large compared to the size of the loop. It can be approximated by calculating the magnitude and direction of the field created by the wire at the center of the loop. Once this is known, the torque on the loop from this applied field can be calculated. The torque on a loop in a uniform magnetic field is given in Equation 24.11.

Solve: (a) The expression for the field created by a wire at the location of the loop is given in Equation 24.1:

$$B = \frac{\mu_0 I}{2\pi r} = \frac{(4\pi \times 10^{-7}\ \text{T} \cdot \text{m/A})(2.0\ \text{A})}{2\pi \times (0.02\ \text{m})} = 2.0 \times 10^{-5}\ \text{T}$$

Using the right-hand rule, the direction of the field will be pointing straight up at the center of the loop, relative to the figure.

To determine the torque on the loop, we need to determine the angle between this field and the dipole moment of the loop. The dipole moment of the loop points directly left relative to Figure P24.41. So the angle between the applied field and the dipole moment of the loop is $90°$.

Using Equation 24.11:

$$\tau = (IA)B\sin\theta = (0.20\ \text{A})(\pi(0.001\ \text{m})^2)(2.0 \times 10^{-5}\ \text{T})\sin(90°) = 1.3 \times 10^{-11}\ \text{N} \cdot \text{m}$$

(b) The dipole will rotate until the angle between the dipole moment and the field is zero, therefore, it will rotate clockwise $90°$. Once it has rotated to its new position, the angle between the dipole moment and the field will be zero, and the loop will feel no torque or net force. At this point, it will be in equilibrium.

P24.47. Prepare: A magnetic field exerts a magnetic force on a length of current carrying wire. We ignore gravitational effects and focus on the magnetic effects. The figure shows a wire in a magnetic field that is directed out of the page. The magnetic force on the wire is therefore to the right and will stretch the springs.

Solve: The direction of the magnetic force is to the right; whereas, the direction of the spring forces is to the left. In static equilibrium, the sum of the forces on the wire is zero:

$$\vec{F}_B + \vec{F}_{\text{spring 1}} + \vec{F}_{\text{spring 2}} = \vec{0}\ \text{N} \Rightarrow ILB + (-k\Delta x) + (-k\Delta x) = 0 \Rightarrow I = \frac{2k\Delta x}{LB} = \frac{2(10\ \text{N/m})(0.01\ \text{m})}{(0.20\ \text{m})(0.5\ \text{T})} = 2.0\ \text{A}$$

P24.49. Prepare: Direct the current so the magnetic field points down.
Solve: (a)

$$B = \frac{\mu_0 NI}{2R} = \frac{(1.26 \times 10^{-6}\ \text{T} \cdot \text{m/A})(100\ \text{turns})(5.0\ \text{A})}{2(3.0\ \text{m})} = 0.105\ \text{mT}$$

which report to two significant figures as 0.11 mT.
(b) Use the righthand rule for forces; with $\vec{v}$ pointing east and $\vec{B}$ pointing down, the force points north.
(c) The magnitude of the force is given by

$$F = |q|vB\sin\alpha = (1.6 \times 10^{-19}\ \text{C})(1.5\ \text{m/s})(0.105\ \text{mT})(1) = 2.5 \times 10^{-23}\ \text{N}$$

Assess: This is a small force, but with a the small mass of the ion, it could still move the ions.

P24.51. Prepare: This question asks us to find the magnetic force between the wires and the magnetic force between the wires and the earth, and to say which is larger. To calculate the magnetic force between the wires we combine two equations to get $F_{\text{wires}} = \dfrac{\mu_0 L I_1 I_2}{2\pi d}$, but we must leave the equation in terms of L, which is not given to us. Next, we calculate the force of the earth on the wires using Equation 24.10, leaving this also in terms of L. We will see that the value of the earth's force on wires is larger than that of the wires on each other.

Solve:

$$F_{\text{wires}} = \frac{\mu_0 L I_1 I_2}{2\pi d} = \frac{(4\pi\times10^{-7}\text{ T}\cdot\text{m/A})(400\text{ A})(400\text{ A})L}{2\pi(5.0\text{ m})}$$

$$F_{\text{wires}} = (0.0064\text{ N/m})L$$

$$F_{\text{earth}} = ILB = (400\text{ A})(5.0\times10^{-5}\text{ T})L = (0.02\text{ N/m})L$$

$$F_{\text{earth}}/F_{\text{wires}} \approx 3$$

The force of the earth's field on the wires is greater.

Assess: The force of the wires on each other would be greater if the two wires were closer enough together.

P24.53. Prepare: As you can see from the figure, the field of the earth, the field of the electromagnet, and the total field form an equilateral triangle. Thus all three sides are the same length. This means that the field of the electromagnet should have a strength of 5.0×10^{-5} T. You can also see from the figure that the field of the electromagnet points south.

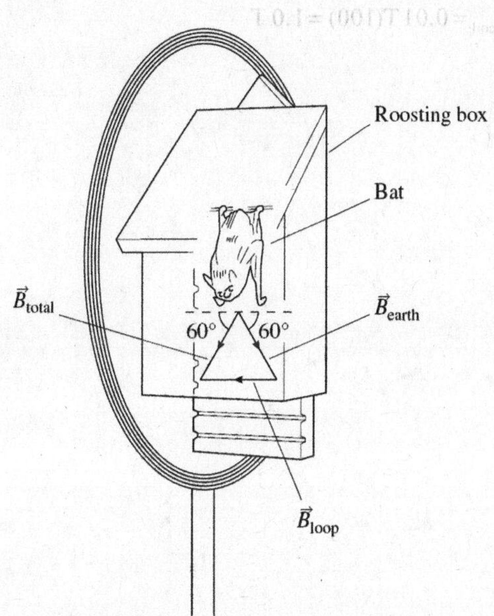

Solve: Since the field of the electromagnet must point south, its axis should be lined up north-south. We find the current by insisting that the total field at the center of the coil be 5.0×10^{-5} T. The field of N loops of wire is given by $B = \dfrac{\mu_0 N I}{2R}$ which we can solve for the current to obtain:

$$I = \frac{2RB}{\mu_0 N} = \frac{(0.50\text{ m})(5.0\times10^{-5}\text{ T})}{(4\pi\times10^{-7}\text{ T}\cdot\text{m/A})(100)} = 0.20\text{ A}$$

Assess: The formula $B_{\text{loop}} = \mu_0 I/(2R)$ only applies close to the center of the loop, so for the experiment to be done accurately, the roosting box should be small compared to the size of the loops of wire. Otherwise, a solenoid might be a better choice since it has a nearly uniform magnetic field in its interior.

P24.55. Prepare: Imagine a piece of metal in the set which has been magnetized. We don't know which way its field points so we impose a new field on the piece by exposing the metal to a strong magnetic field in a known direction. But we still have a magnetized piece of metal. We have to get rid of this new field. So we rapidly reverse the direction of the applied field but we don't let it become as intense in the opposite direction as it was in the original direction. For example, the field might be 1.0 T in one direction and then we might rapidly change it to 0.90 T in the opposite direction. The magnetic moments flip back and forth in response to the external field but the induced field is steadily decreasing since we decrease the applied field each time we change its direction. Eventually, the piece of metal will have an immeasurably small field and will no longer distort the picture on the screen.

Assess: Since you cannot make a piece of metal have zero magnetic dipole by exposing it to zero applied field (which would be, essentially, doing nothing), we have to use this bizarre direction reversing approach in which the applied field is gradually reduced to zero. Some television and computer monitors undergo this process, known as degaussing, every time they are turned on. This is why you may see the image shake back and forth for a short time after turning the device on.

P24.57. Prepare: We must start with the normal equation for a coil, Equation 24.4. Once we've calculated the magnetic field, it is simply a matter of multiplying it by a factor of 100.
Solve:

$$B_{\text{coil}} = \mu_0 \left(\frac{N}{L} \right) I = \frac{(4\pi \times 10^{-7}\,\text{T} \cdot \text{m/A})(240)(0.60\,\text{A})}{0.018\,\text{m}}$$

$$B_{\text{coil}} = 0.01\,\text{T}$$

$$B_{\text{in coil}} = 0.01\,\text{T}(100) = 1.0\,\text{T}$$

25

EM INDUCTION AND EM WAVES

Q25.1. Reason: With the assumption that the loop is stationary, no current will be induced in the loop because the magnetic flux through it is not *changing*.
Assess: It doesn't matter how strong the magnetic field is (or how big the loop is), if the field through the loop isn't changing then there won't be any current induced in it. Lenz's law tells us this.

Q25.3. Reason: In each case we apply Lenz's law: "The direction of the induced current is such that the induced magnetic field opposes the *change* in the flux."
(a) The magnetic field is out of the page and the flux through the rectangular loop is increasing because the loop is growing. A current will be induced that will create a magnetic field to oppose that change. The induced magnetic field must be into the page; this requires (by the right-hand rule) that the induced current be clockwise.
(b) The magnetic field is into the page and the flux through the rectangular loop is increasing because the loop is growing. A current will be induced that will create a magnetic field to oppose that change. The induced magnetic field must be out of the page; this requires (by the right-hand rule) that the induced current be counterclockwise.
(c) The magnetic field is parallel to the plane of the page and so the flux through the rectangular loop is zero and not changing. Since the flux through the loop isn't changing then the induced current is zero.
(d) The magnetic field is out of the page and the flux through the triangular loop is decreasing because the loop is shrinking. A current will be induced that will create a magnetic field to oppose that change. The induced magnetic field must be out of the page; this requires (by the right-hand rule) that the induced current be counterclockwise.
(e) The magnetic field is into the page but the flux through the rectangular loop is constant because the loop is not growing or shrinking as the two rails move at the same velocity. Since the flux through the loop isn't changing then the induced current is zero.
(f) The magnetic field is into the page but the flux through the rectangular loop is constant because the moving wire is outside the region of the magnetic field; as the loop grows it doesn't enclose any more flux. Since the flux through the loop isn't changing then the induced current is zero.
Assess: Flux though a loop of wire can change for various reasons, either because the area of the loop changes, or the strength of the field changes, or the orientation of the loop changes. Only if the flux through the loop changes is a current induced.

Q25.7. Reason: The magnetic field lines generated by a long current-carrying wire are concentric circles around the wire. The strength of that field changes continuously in the AC power lines.
If you orient your loop so its axis is parallel to the power line then you won't ever get any flux through it and certainly no changing flux through it. So that is a bad idea.
To get the most flux through your loop you want to place it close to the power line (where the field is strongest) and in an orientation so the plane of the loop also contains the power line. Not only will this get you the greatest value of the flux, but in this case it also produces the greatest *change* in the flux and therefore the greatest induced emf in your loop as the field goes from a maximum to zero and then to a maximum in the other direction.

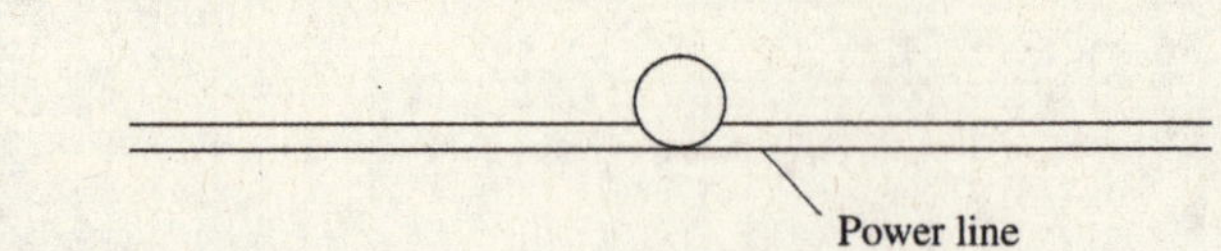

Assess: We also don't want the loop to be flat on the power line (with two opposite points of the loop touching the line) because some field lines would be going through the loop one way and other lines going through the opposite way leaving you with very little flux through the loop.

Q25.11. Reason: (a) As you shove the coil into the region of the magnetic field, the magnetic flux through the coil will be increasing. This changing magnetic flux will create an induced clockwise current (see the figure below).
(b) This induced current interacts with the existing magnetic field in such a manner as to oppose what caused it. Since the induced current was caused by you pushing the coil into the region of the magnetic field, as the induced current interacts with the existing magnetic field the resulting force will oppose its motion into the region of the field. The figure below shows the magnet, magnetic field, coil, induced current in the coil, the force you are exerting, and the subsequent magnetic force. As shown in the figure, you need to push the loop in against a repulsive force.

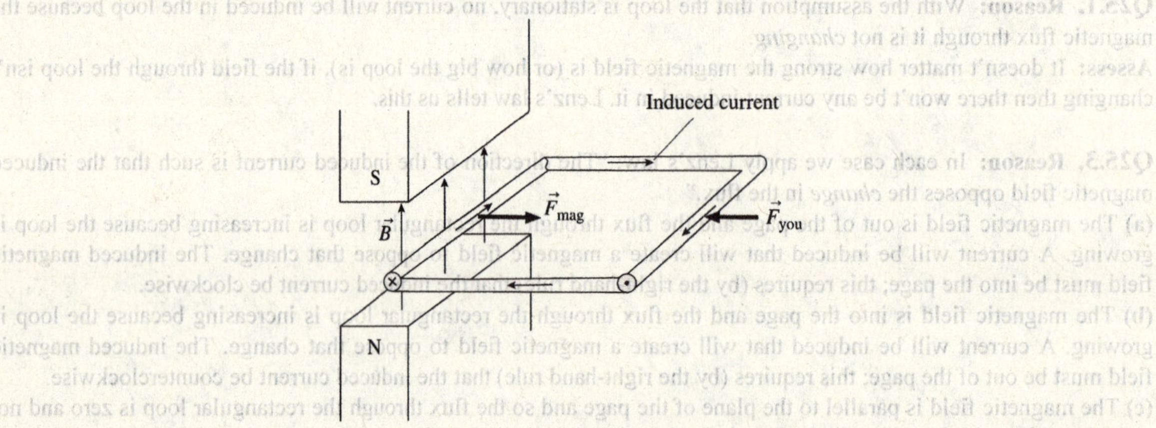

Assess: This problem is an excellent example of Lenz's law. As the coil is pushed into the region of the field, there is an increasing upward magnetic flux. The induced current creates an induced magnet flux that is directed downward.

Q25.13. Reason: When the current in the left ring is positive the magnetic field lines come up out of the left ring and some of them then go down through the right ring to connect back up again from below in the left ring. So when the field points up in the left ring it points down in the right ring. When a positive current in the left ring is increasing the field pointing down in the right ring is increasing; an induced current in the right ring will be in the direction to oppose the increase, i.e., the induced current will be counterclockwise (positive). Combine the equations to show that the induced current in the right loop depends on the slope of the graph of the current in the left loop.

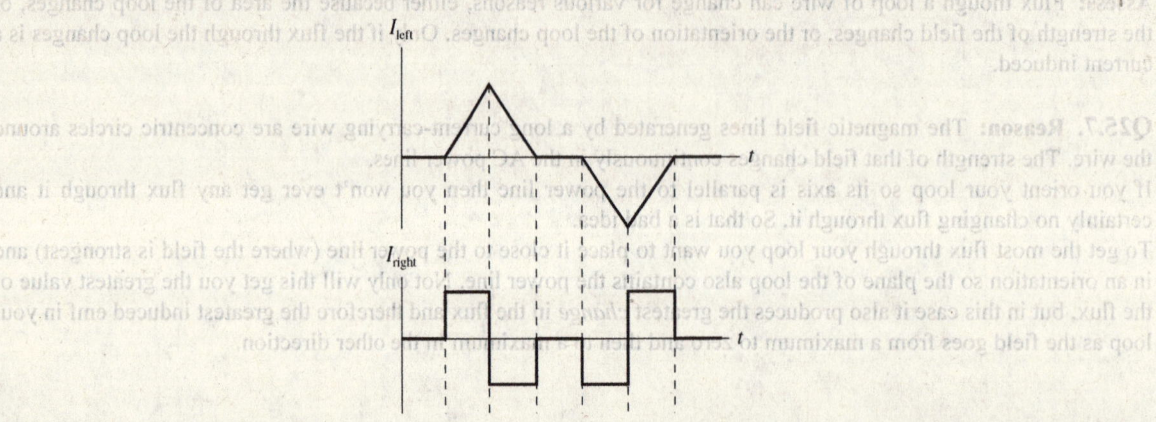

Assess: The induced current in the right loop depends on the *change* in the field through it, which directly related to the *change* in the current in the left loop, not the value of the current in the left loop.

Q25.15. Reason: (a) As the north magnetic pole of the magnet approaches the coil, there is an increasing upward magnetic flux through the coil. The induced current must be in such a direction as to oppose this increase in magnetic flux. If the induced current is clockwise (as viewed from above), the induced magnetic field will be downward. As the coil is pushed into the field, the upward magnetic flux increases, the induced clockwise current increases, the induced magnetic field increases and the induced magnetic flux directed in the downward direction increases.

(b) When the magnet is being pushed toward the coil, the magnetic situation at the site of the coil is as shown in the following figure. The magnetic field at the site of the coil may be resolved into a vertical and horizontal component.

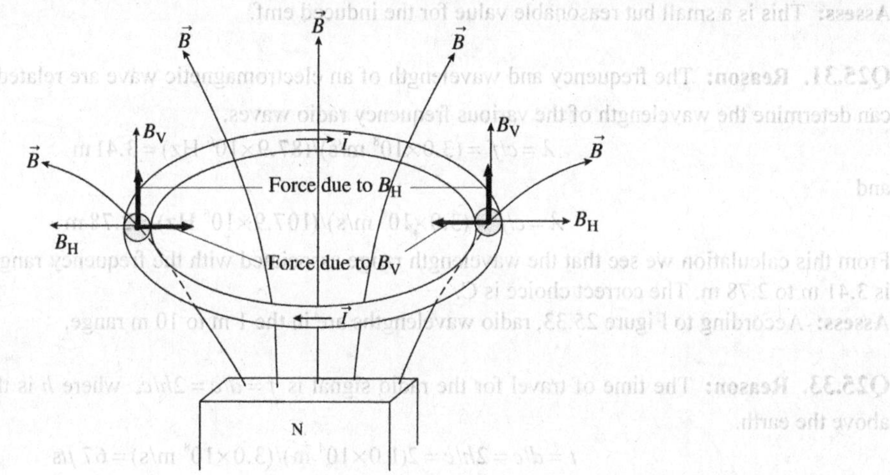

Recall that the current is clockwise as indicated by the tail of an arrow representing the current at the left and the head of an arrow at the right. Notice that as the induced current interacts with the magnetic field of the approaching magnet, the vertical component of the magnetic field results in forces that try to collapse the coil and the horizontal component of the magnetic field results in forces that try to push the coil vertically upward. **(c)** By Newton's third law, if the magnet exerts an upward force on the loop, the loop must exert a downward force on the magnet.
Assess: If the magnetic field associated with the magnet is strong enough and magnet is thrust toward the coil fast enough, the resulting induced forces will support the coil.

Q25.19. Reason: The light we are interested in blocking is the light reflected off water and other horizontal surfaces. This light is partially polarized parallel to the plane of the reflecting surface, hence the electric field vector we wish to block is oscillating parallel to the horizontal surface. If the long axis of the polarizing molecules is also parallel to the horizontal surface, the electrons in the molecules will be forced into oscillations along the long molecules by the electric field. During this process, the polarizing coating on the surface of the sunglasses absorbs the energy of the electric field and as a result the electric field does not pass through the sunglasses.
Assess: Polarized sunglasses are a good example of better living through a combination of chemistry and physics.

Q25.21. Reason: (a) As the oscillating magnetic field of the radio wave goes through the coil, it induces a little current in the coil because the flux through it is changing. This small current is amplified and then fed to the speakers.
(b) The coil and ferrite bar need to be horizontal since the magnetic field will be oscillating in a horizontal plane (perpendicular to the vertical plane of the electric field). If the station is north of you then the magnetic field will oscillate in an east-west direction, and that is the way you should orient your ferrite bar to detect the largest changes in flux.
Assess: Look at some household AM radios. The ferrite bar on the back *is* horizontal, but whether it points east-west is up to you and where your favorite station broadcasts from.

Q25.27. Reason: The figure indicates that the magnetic field is uniform and we are going to assume that the coil is rotating at a constant angular frequency. As a result the magnetic flux in the coil is constant. Since the magnetic flux in the coil is not changing, no electric current will be induced. The correct choice is C.

Assess: A changing magnetic flux is needed in order to induce an electric current.

Q25.29. Reason: The magnetic flux through the coil is $\Phi = BA = B(L^2/2)$. As the magnetic field changes the magnetic flux changes and the induced emf, which is the rate of change of the magnetic flux, is determined by

$$\mathcal{E} = \frac{\Delta\Phi}{\Delta t} = \frac{\Delta(BA)}{\Delta t} = \frac{\Delta B}{\Delta t}A = \frac{\Delta B}{\Delta t}(L^2/2) = \frac{0.2\text{ T}}{5.0\times10^{-2}\text{ s}}(0.2\text{ m})^2/2 = 0.08\text{ V}$$

The correct choice is A.

Assess: This is a small but reasonable value for the induced emf.

Q25.31. Reason: The frequency and wavelength of an electromagnetic wave are related by $c = f\lambda$. Using this we can determine the wavelength of the various frequency radio waves.

$$\lambda = c/f = (3.0\times10^8\text{ m/s})/(87.9\times10^6\text{ Hz}) = 3.41\text{ m}$$

and

$$\lambda = c/f = (3.0\times10^8\text{ m/s})/(107.9\times10^6\text{ Hz}) = 2.78\text{ m}$$

From this calculation we see that the wavelength range associated with the frequency range 87.9 MHz to 107.9 MHz is 3.41 m to 2.78 m. The correct choice is C.

Assess: According to Figure 25.33, radio wavelengths are in the 1 m to 10 m range.

Q25.33. Reason: The time of travel for the radio signal is $t = d/c = 2h/c$, where h is the height of the spacecraft above the earth.

$$t = d/c = 2h/c = 2(1.0\times10^4\text{ m})/(3.0\times10^8\text{ m/s}) = 67\ \mu\text{s}$$

The correct choice is D.

Assess: This is a reasonable amount of time for the signal to travel this distance at the speed of light.

Problem

P25.3. Prepare: The wire is pulled with a constant force in a magnetic field. This results in a motional emf and produces a current in the circuit. From energy conservation, the mechanical power provided by the puller must appear as electrical power in the circuit.

Solve: (a) Using Equation 25.6,

$$P = F_{\text{pull}}v \Rightarrow v = \frac{P}{F_{\text{pull}}} = \frac{4.0\text{ W}}{1.0\text{ N}} = 4.0\text{ m/s}$$

(b) Using Equation 25.6 again,

$$P = \frac{v^2 l^2 B^2}{R} \Rightarrow B = \sqrt{\frac{RF_{\text{pull}}}{vl^2}} = \sqrt{\frac{(0.20\ \Omega)(1.0\text{ N})}{(4.0\text{ m/s})(0.10\text{ m})^2}} = 2.2\text{ T}$$

Assess: This is a reasonable field for the circumstances given.

P25.5. Prepare: We neglect air resistance for this case. The terminal speed will be achieved when the downward force of gravity on the bar is equal in magnitude to the upward force due to the induced current in the bar in the horizontal magnetic field. We'll use Equation 25.5 for the force needed to produce a constant speed.

Known
$l = 0.12$ m
$B = 0.060$ T
$R = 1.0\ \Omega$
$m = 0.050$ kg

Find
v

We use the given mass $m = 0.050$ kg to compute the gravitational force $w = mg = (0.050\ \text{kg})(9.8\ \text{m/s}^2)$.

Solve: See Equation 25.5. For terminal speed

$$F_{\text{grav}} = F_{\text{mag}}$$

$$mg = IlB$$

$$mg = \left(\frac{vlB}{R}\right)lB$$

$$mg = \frac{vl^2B^2}{R}$$

Solve for v.

$$v = \frac{mgR}{l^2B^2} = \frac{(0.050\ \text{kg})(9.8\ \text{m/s}^2)(1.0\ \Omega)}{(0.12\ \text{m})^2(0.060\ \text{T})^2} = 9500\ \text{m/s}$$

Assess: The answer is quite large, but that is because most real magnetic fields are quite weak (as is this one). If we hadn't neglected air resistance the bar would have reached terminal speed due to air resistance long before it does due to the magnetic force.

P25.9. Prepare: The direction of magnetic field is shown as left to right. The direction of the vector representing the area of the coil is perpendicular to the coil. The angle θ is the angle between the direction of the magnetic field $\vec{B}$ and the area $\vec{A}$. The magnetic flux is obtained by $\Phi = BA\cos\theta$.

Solve: An expression for the magnetic flux for any angle θ is as follows:

$$\Phi = BA\cos\theta = B\pi r^2 \cos\theta = (0.05T)\pi(0.050m)^2 \cos\theta = (3.93\times10^{-4}\ \text{Wb})\cos\theta$$

For the angle $\theta = 0°$ obtain $\Phi = 3.93\times10^{-4}$ Wb $\approx 3.9\times10^{-4}$ Wb

$\theta = 30°$ obtain $\Phi = 3.40\times10^{-4}$ Wb $\approx 3.4\times10^{-4}$ Wb

$\theta = 60°$ obtain $\Phi = 1.97\times10^{-4}$ Wb $\approx 2.0\times10^{-4}$ Wb

$\theta = 90°$ obtain $\Phi = 0$ WB

Assess: Given the strength of the magnetic field and the area of the coil, these are reasonable values for the magnetic flux.

P25.11. Prepare: Consider the solenoid to be long so the field is constant inside and zero outside. The field of a solenoid is along the axis. The flux through the loop is only nonzero inside the solenoid. Since the loop completely surrounds the solenoid, the total flux through the loop will be the same in both the perpendicular and tilted cases.

Solve: (a) The field is constant inside the solenoid so we will use Equation 25.9. Take $\vec{A}$ to be in the same direction as the field. The magnetic flux is

$$\Phi = A_{\text{loop}}B_{\text{loop}}\cos\theta = A_{\text{sol}}B_{\text{sol}}\cos\theta = \pi r_{\text{sol}}^2 B_{\text{sol}}\cos\theta = \pi(0.010\ \text{m})^2(0.20\ \text{T}) = 6.3\times10^{-5}\ \text{Wb}$$

(b) When the loop is tilted, the component of $\vec{B}$ in the direction of $\vec{A}$ is less, but the effective area of the loop surface through which the magnetic field lines cross is increased by the same factor, so the answer is the same, 6.3×10^{-5} Wb .

Assess: Because the field created is along the length of the solenoid, we used $\theta = 0°$.

P25.13. Prepare: Assume the plane of the loop is perpendicular to the field direction. The flux is due to the field through the area of the triangle. Only the left half gives a contribution as the field strength is zero on the right half.

Solve: (a) The flux given by Equation 25.9 is $\Phi = AB\cos\theta$. Take $\vec{A}$ to be into the page, perpendicular to the triangle, and thus parallel to $\vec{B}$. In this case $\Phi = AB$ where A is the area of half of the triangle. This smaller triangle has a base of 10 cm and height $20\sin 60° \text{ cm} = 17.32$ cm. Thus,

$$\Phi = AB = \frac{1}{2} = (0.10 \text{ m})(0.1732 \text{ m}) \times 0.1 \text{ T} = 8.7 \times 10^{-4} \text{ Wb}$$

(b) The flux is directed into the loop. According to Lenz's law, the induced current will try to *prevent the decrease* of flux. To do this, the field of the induced current will have to point into this loop. This requires the induced current to flow *clockwise*.

P25.15. Prepare: Assume the field strength is changing at a constant rate. The changing field produces a changing flux in the coil and there will be a corresponding induced emf and current. We will use Equations 25.9 and 25.12. When $\vec{B}$ is parallel to $\vec{A}$, $\Phi = BA\cos 0° = BA$.

$$N = 1000 \text{ turns}$$

Solve: The induced emf of the coil is

$$\varepsilon = N\left|\frac{\Delta\Phi}{\Delta t}\right| = N\left|\frac{\Delta(AB)}{\Delta t}\right| = NA\left|\frac{\Delta B}{\Delta t}\right| = N\pi r^2\left|\frac{\Delta B}{\Delta t}\right| = (10^3)\pi(0.01 \text{ m})^2\left(\frac{0.10 \text{ T}}{10\times10^{-3} \text{ s}}\right) = 3.1 \text{ V}$$

Assess: This seems to be a reasonable emf as there are many turns.

P25.21. Prepare: The induced emf is determined by $\varepsilon_{\text{induced}} = \Delta\Phi/\Delta t = A(\Delta B/\Delta t)$. The induced emf is in the same direction as the applied voltage (the 9.0 volts from the battery), so the net emf is $\varepsilon_{\text{net}} = \varepsilon_{\text{applied}} + \varepsilon_{\text{induced}}$. The net emf and current are related by $\varepsilon_{\text{net}} = IR$. Combining these expressions and solving for the current we obtain: $I = [\varepsilon_{\text{applied}} + A(\Delta B/\Delta t)]/R$.

Solve: The current in the circuit and hence the resistor is

$$I = \frac{\left[\varepsilon_{\text{applied}} + A\left(\dfrac{\Delta B}{\Delta t}\right)\right]}{R} = \frac{\left[9.0 \text{ V} + (400\times10^{-4} \text{ m}^2)\left(\dfrac{0.50 \text{ T}}{10\times10^{-3}\text{ s}}\right)\right]}{20 \text{ } \Omega} = 0.55 \text{ A}$$

Assess: The current in the circuit is decreased slightly due to the fact that the induced emf opposes the applied voltage.

P25.23. Prepare: Combine the equation for the induced emf with Ohm's law. Assume the axis of the coil is parallel to the field so the cosine of the angle is 1. The area of the coil is $A = \pi R^2 = \pi\left(\dfrac{C}{2\pi}\right)^2 = \pi\left(\dfrac{0.20 \text{ m}}{2\pi}\right)^2 = 0.00318 \text{ m}^2$.

Solve:

$$I = \frac{\mathcal{E}}{R} = \frac{\dfrac{\Delta\Phi}{\Delta t}}{R} = \frac{\dfrac{\Delta(AB\cos\phi)}{\Delta t}}{R} = \frac{\dfrac{A\Delta B}{\Delta t}}{R} = \frac{\dfrac{(0.00318\ \text{m}^2)(0.55\ \text{T})}{(15\times10^{-3}\,\text{s})}}{0.12\ \Omega} = 0.97\ \text{A}$$

Assess: This is a reasonable current.

P25.25. Prepare: The electric and magnetic field amplitudes of an electromagnetic wave are related as $E_0 = cB_0$.
Solve: The electric field amplitude of the electromagnetic wave is

$$E_0 = cB_0 = (3.0\times10^8\ \text{m/s})(2.0\times10^{-3}\ \text{T}) = 6.0\times10^5\ \text{V/m}$$

Assess: Because the magnetic field amplitude is much larger than the earth's magnetic field, we expected a large electric field amplitude.

P25.29. Prepare: The intensity of the microwaves is related to the amplitude of the oscillating electric field by $I = c\mathcal{E}_0 E_0^2/2$ and to the power and cross-sectional area of the beam by $I = P/A$. The amplitude of the oscillating magnetic field is related to the amplitude of the oscillating electric field by $B_0 = E_0/c$.
Solve: The amplitude of the oscillating electric field is

$$E_0 = (2I/(c\mathcal{E}_0))^{1/2} = [2P/(\pi r^2 c\mathcal{E}_0)]^{1/2}$$
$$[2(10^{-3}\ \text{W})/\pi(5.0\times10^{-4}\ \text{m})^2(3.0\times10^8\ \text{m/s})(8.85\times10^{-12}\ \text{C}^2/(\text{N}\cdot\text{m}^2))]^{1/2} = 980\ \text{V/m}$$

The amplitude of the oscillating magnetic field is

$$B_0 = E_0/c = (980\ \text{V/m})/(3.0\times10^8\ \text{m/s}) = 3.3\times10^{-6}\ \text{T}$$

Assess: These are reasonable values for the amplitude of the oscillating electric and magnetic fields of microwaves.

P25.31. Prepare: The energy transported per second by the wave is 10 W or 10 J/s. This energy is carried uniformly in all directions. Equation 25.18 connects power of a wave that impinges on an area A with the wave's electric field amplitude.
Solve: From Equation 25.18, the light intensity is

$$I = \frac{P}{4\pi r^2} = \frac{c\mathcal{E}_0}{2}E_0^2 \Rightarrow r = \sqrt{\frac{2P}{4\pi c\mathcal{E}_0 E_0^2}} = \sqrt{\frac{2(10\times10^{-3}\ \text{W})}{4\pi(3.0\times10^8\ \text{m/s})(8.85\times10^{-12}\ \text{C}^2/(\text{N}\cdot\text{m}^2))(0.010\ \text{V/m})^2}} = 77\ \text{m}$$

Assess: Greater distances result in smaller electric fields. This is exactly what we expect.

P25.35. Prepare: We will use Malus's law for the polarized light. From Equation 25.21 the relationship between the incident and the transmitted polarized light is $I_{\text{transmitted}} = I_{\text{incident}}\cos^2\theta$, where θ is the angle between the electric field and the axis of the filter. In our case θ will be the angle between the filter axis and the plane of polarization.
Solve: From Equation 25.21 the intensity is
(a) $I_{\text{transmitted}} = I_{\text{incident}}\cos^2\theta = (10\ \text{W/m}^2)\cos^2(0°) = 10\ \text{W/m}^2$
(b) $I_{\text{transmitted}} = I_{\text{incident}}\cos^2\theta = (10\ \text{W/m}^2)\cos^2(30°) = 7.5\ \text{W/m}^2$
(c) $I_{\text{transmitted}} = I_{\text{incident}}\cos^2\theta = (10\ \text{W/m}^2)\cos^2(45°) = 5.0\ \text{W/m}^2$
(d) $I_{\text{transmitted}} = I_{\text{incident}}\cos^2\theta = (10\ \text{W/m}^2)\cos^2(60°) = 2.5\ \text{W/m}^2$
(e) $I_{\text{transmitted}} = I_{\text{incident}}\cos^2\theta = (10\ \text{W/m}^2)\cos^2(90°) = 0\ \text{W/m}^2$

Assess: The final intensity must be between $0\ \text{W/m}^2$ and $10\ \text{W/m}^2$ and all values are in this range.

P25.37. Prepare: We will use Malus's law for the polarized light. From Equation 25.21 the relationship between the incident and the transmitted polarized light is $I_{\text{transmitted}} = I_0\cos^2\theta$, where θ is the angle between the electric field and the axis of the filter.

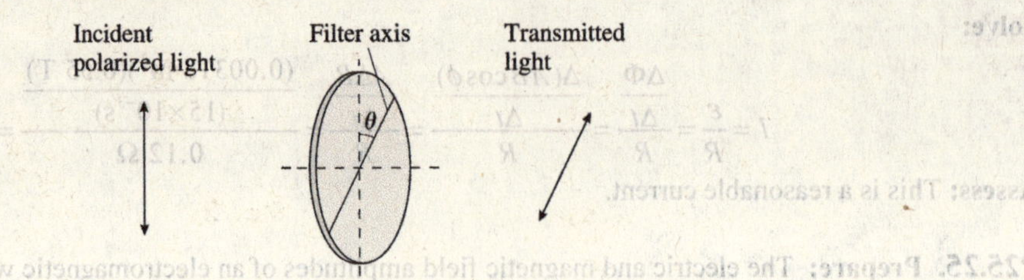

Solve: From Equation 25.21,

$$0.25\, I_0 = I_0 \cos^2 \theta \Rightarrow \cos \theta = 0.50 \Rightarrow \theta = 60°$$

Assess: Note that θ is the angle between the electric field and the axis of the filter.

P25.41. Prepare: We will use the photon model of light as given by Equation 25.22. We will also use $f = c/\lambda$ in Equation 25.22.

Solve: The energy of the x-ray photon is

$$E = hf = h\left(\frac{c}{\lambda}\right) = (6.63\times10^{-34}\ \text{J}\cdot\text{s})\left(\frac{3.0\times10^{8}\ \text{m/s}}{1.0\times10^{-9}\ \text{m}}\right) = 1.99\times10^{-16}\ \text{J}$$

In eV, the energy is $(1.99\times10^{-16}\ \text{J})\left(\frac{1\,\text{eV}}{1.6\times10^{-19}\,\text{J}}\right) = 1200\ \text{eV}$.

Assess: This is a small amount of energy, but it is larger than the energy of a photon in the visible wavelength.

P25.45. Prepare: The relationship between energy and wavelength for a photon is $E = hc/\lambda$.

Solve: The energy of the photons is $E = hc/\lambda = (6.63\times10^{-34}\ \text{J}\cdot\text{s})(3.0\times10^{8}\ \text{m/s})/(9.5\times10^{-6}\ \text{m}) = 2.1\times10^{-20}\ \text{J}$.

This is equal to 1.3 eV.

Assess: We expect the energy of a single photon of this wavelength to be small.

P25.47. Prepare: We follow somewhat the strategy outlined in Example 25.9. Along the way we'll use these relationships: intensity = power/area and energy = power×time.

We'll first find the energy per photon for the given wavelength from $E = hf$ where $f = c/\lambda$.

We are given $\lambda = 680\times10^{-9}$ m, $A = 1.00\ \text{m}^2$, $\Delta t = 1.00$ s, and the intensity $I = 1370\ \text{W/m}^2$.

Solve:

$$f = \frac{c}{\lambda} = \frac{3.00\times10^{8}\ \text{m/s}}{680\times10^{-9}\ \text{m}} = 4.412\times10^{14}\ \text{Hz}$$

The energy per photon is

$$E_{\text{photon}} = hf = (6.63\times10^{-34}\ \text{J}\cdot\text{s})(4.412\times10^{14}\ \text{Hz}) = 2.925\times10^{-19}\ \text{J}$$

Now we estimate the number of photons by applying the relationships.

$$N \approx \frac{\text{total energy}}{\text{energy/photon}} = \frac{\text{power}\times\text{time}}{\text{energy/photon}} = \frac{(\text{intensity area})\times\text{time}}{\text{energy/photon}}$$

$$N \approx \frac{IA\Delta t}{E/\text{photon}} = \frac{(1370\ \text{W/m}^2)(1.00\ \text{m}^2)(1.00\ \text{s})}{2.925\times10^{-19}\ \text{J/photon}} = 4.68\times10^{21}\ \text{photons}$$

Assess: Don't confuse the usage of I as intensity here with its usage elsewhere for current.
The photons emitted by the sun will span a range of energies, because the light spans a range of wavelengths, but the average photon energy is assumed to be 680 nm.

P25.51. Prepare: Wein's law, given in Equation 25.24, tells us the relationship between the temperature and the peak wavelength, although the answer will be in kelvin and we'll then convert to °C.

$$\lambda_{\text{peak}}\ (\text{in nm}) = \frac{2.9\times10^{6}\ \text{nm}\cdot\text{K}}{T}$$

We are given $\lambda_{\text{peak}} = 1200$ nm.

Solve: Solve the equation for T.

$$T = \frac{2.9 \times 10^6 \, \text{nm} \cdot \text{K}}{\lambda_{\text{peak}}} = \frac{2.9 \times 10^6 \, \text{nm} \cdot \text{K}}{1200 \, \text{nm}} = 2417 \, \text{K}$$

Now we must convert the answer in kelvin to $^\circ$C by subtracting 273: $2417 \, \text{K} - 273^\circ = 2144 \,^\circ\text{C}$.
This should be rounded to $2100 \,^\circ\text{C}$.

Assess: Regular tungsten filaments are usually a bit hotter than this, but we are in the right ballpark.

P25.53. Prepare: The wavelength and the maximum intensity are related by λ_{peak} (in nm) $= (2.9 \times 10^6 \, \text{nm} \cdot \text{K})/T$
(in K). According to the Stefan-Boltzmann law $(\Delta Q/\Delta t = e\sigma AT^4)$, the thermal radiation is proportional to the
temperature raised to the fourth power (T^4).

Solve: (a) The temperature for the given wavelengths is

$$T_{1800 \, \text{nm}} = (2.9 \times 10^6 \, \text{nm} \cdot \text{K})/\lambda = (2.9 \times 10^6 \, \text{nm} \cdot \text{K})/(1.8 \times 10^3 \, \text{nm}) = 1.6 \times 10^3 \, \text{K}$$

and

$$T_{1600 \, \text{nm}} = (2.9 \times 10^6 \, \text{nm} \cdot \text{K})/\lambda = (2.9 \times 10^6 \, \text{nm} \cdot \text{K})/(1.6 \times 10^3 \, \text{nm}) = 1.8 \times 10^3 \, \text{K}$$

The resulting temperature change is $\Delta T = T_{1600 \, \text{nm}} - T_{1800 \, \text{nm}} = 200 \, \text{K} = 200 \,^\circ\text{C}$.
The positive sign informs us that the temperature increases.
(b) The energy radiated for each wavelength case is

$$(\Delta Q/\Delta t)_{1800 \, \text{nm}} = e\sigma A T_{1800 \, \text{nm}}^4 = e\sigma A(1.6 \times 10^3 \, \text{K})^4 = e\sigma A(6.5 \times 10^{12} \, \text{K}^4)$$

and

$$(\Delta Q/\Delta t)_{1600 \, \text{nm}} = e\sigma A T_{1600 \, \text{nm}}^4 = e\sigma A(1.8 \times 10^3 \, \text{K})^4 = e\sigma A(10.5 \times 10^{12} \, \text{K}^4)$$

$$\frac{(\Delta Q/\Delta t)_{1600 \, \text{nm}}}{(\Delta Q/\Delta t)_{1800 \, \text{nm}}} = \frac{e\sigma A(10.5 \times 10^{12} \, \text{K}^4)}{e\sigma A(6.5 \times 10^{12} \, \text{K}^4)} = 1.6$$

This informs us that as the peak wavelength is decreased from 1800 nm to 1600 nm, the temperature increases by
$200 \,^\circ\text{C}$ and the thermal radiation increases by a factor of 1.6.

Assess: Looking at Figure 25.39, notice that as the peak wavelength goes down the temperature goes up and the
intensity of the radiation increases. This is consistent with our calculation.

P25.55. Prepare: For both cases we will use $E_{\text{photon}} = hf$ to find the frequency, then $c = \lambda f$ to arrive at the
wavelength.
In the second case the energy of the photons is $E = (2.5)(25 \, \text{keV}) = 62.5 \, \text{keV}$.

Solve: For the 25 KeV photons:

$$f = \frac{E}{h} = \frac{25000 \, \text{eV}}{6.63 \times 10^{-34} \, \text{J} \cdot \text{s}} \left(\frac{1.6 \times 10^{-19} \, \text{J}}{1 \, \text{eV}} \right) = 6.0 \times 10^{18} \, \text{Hz}$$

$$\lambda = \frac{c}{f} = \frac{3.0 \times 10^8 \, \text{m/s}}{6.0 \times 10^{18} \, \text{Hz}} = 5.0 \times 10^{-11} \, \text{m} = 50 \, \text{pm}$$

For the 62.5 keV photons:

$$f = \frac{E}{h} = \frac{62500 \, \text{eV}}{6.63 \times 10^{-34} \, \text{J} \cdot \text{s}} \left(\frac{1.6 \times 10^{-19} \, \text{J}}{1 \, \text{eV}} \right) = 1.5 \times 10^{19} \, \text{Hz}$$

$$\lambda = \frac{c}{f} = \frac{3.0 \times 10^8 \, \text{m/s}}{1.5 \times 10^{19} \, \text{Hz}} = 2.0 \times 10^{-11} \, \text{m} = 20 \, \text{pm}$$

Assess: We see that the more energetic photons have a shorter wavelength, as we expected. These wavelengths are
in the x-ray range.

P25.59. Prepare: We will assume that B changes uniformly with time. The magnetic strength is changing so the
flux is changing and this will create an induced emf. The magnetic field is at an angle $\theta = 45^\circ$ to the normal to the
plane of the coils.

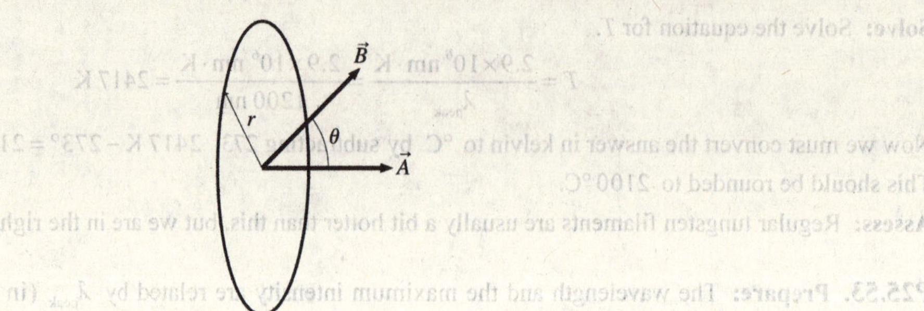

Solve: The flux for a single loop of the coil is $\Phi = BA\cos\theta$. The radius and angle don't change with time, but B does. According to Faraday's law, the induced emf is

$$\mathcal{E} = N\left|\frac{\Delta\Phi}{\Delta t}\right| = NA\cos\theta\left|\frac{\Delta B}{\Delta t}\right| = N\pi r^2 \cos\theta\left|\frac{\Delta B}{\Delta t}\right|$$

$$= 25\pi(0.050\text{ m})^2(\cos 45°)\left|\frac{0.20\text{ T} - 0.80\text{ T}}{2.0\text{ s}}\right| = 4.17\times10^{-2}\text{ V} = 42\text{ mV}$$

Assess: This is a reasonable induced emf for a 25-turn coil and a rather fast decrease in the magnetic field.

P25.63. Prepare: We will assume that the field is uniform in space over the loop. The moving slide wire in a magnetic field develops a motional emf and a corresponding current in the wire and the rails. The current through the resistor will cause energy dissipation and subsequent warming.

Solve: **(a)** The induced emf depends on the changing flux, not on where the resistance in the loop is located. From Equation 25.3, we have $\mathcal{E}_{\text{loop}} = vlB = (10\text{ m/s})(0.20\text{ m})(0.10\text{ T}) = 0.20\text{ V}$.

The total resistance around the loop is due entirely to the carbon resistor, so the induced current is $I = \mathcal{E}_{\text{loop}}/R = (0.20\text{ V})/(1.0\ \Omega) = 0.20\text{ A}$.

(b) To move with a constant velocity the acceleration must be zero, so the pulling force must balance the retarding magnetic force on the induced current in the slide wire. Thus from Equation 25.5, $F_{\text{pull}} = F_{\text{mag}} = IlB = (0.20\text{ A})$ $(0.20\text{ m})(0.1\text{ T}) = 4.0\times10^{-3}\text{ N}$.

(c) The current in the resistor will result in power being dissipated. The power is $P = I^2R = (0.20\text{ A})^2(1\ \Omega) = 0.040\text{ W} = 0.040\text{ J/s}$.

During a 10-second period $Q = 0.40\text{ J}$ of energy is dissipated by the current, increasing the internal energy of the carbon resistor and raising its temperature. This is, effectively, a heat source. The heat is related to the temperature rise by $Q = mc\Delta T$, where c is the specific heat of carbon. Thus,

$$\Delta T = \frac{Q}{mc} = \frac{0.40\text{ J}}{(5.0\times10^{-5}\text{ kg})(710\text{ J/(kg °C)})} = 11\text{ K} = 11\text{ °C}$$

Assess: This is a significant change in the temperature of the resistor.

P25.65. Prepare: The moving wire will have a motional emf that produces a current in the loop. We will assume there is no resistance in the rails, and if there is any resistance, it is accounted for by the resistor.

Solve: (a) At constant velocity the external pushing force is balanced by the magnetic force, so using Equation 25.5

$$F_{push} = F_{mag} = IlB = \frac{\mathcal{E}}{R}lB = \left(\frac{Blv}{R}\right)lB = \frac{B^2l^2v}{R} = \frac{(0.50\text{ T})^2(0.10\text{ m})^2(0.50\text{ m/s})}{2.0\,\Omega} = 6.25\times10^{-4}\text{ N} = 6.3\times10^{-4}\text{ N}$$

(b) The power is $P = Fv = (6.25\times10^{-4}\text{ N})(0.50\text{ m/s}) = 3.1\times10^{-4}\text{ W}$.

(c) The flux is out of the page and decreasing and the induced current/field will oppose the change. The induced field must have a flux that is out of the page so the current will be *counterclockwise*.
The magnitude of the current using Equation 25.4 is

$$I = \left(\frac{Blv}{R}\right) = \frac{(0.50\text{ T})(0.10\text{ m})(0.50\text{ m/s})}{2.0\,\Omega} = 1.25\times10^{-2}\text{ A} = 1.3\times10^{-2}\text{ A}$$

(d) The power is $P = I^2R = (1.25\times10^{-2}\text{ A})^2(2.0\,\Omega) = 3.1\times10^{-4}\text{ W}$.

Assess: From energy conservation we see that the mechanical energy put in by the pushing force shows up as electrical energy in the resistor.

P25.69. Prepare: Intensity is related to power and area by $I = P/A$. Intensity is also related to the amplitude of the oscillating electric field by $I = c\varepsilon_0 E_0^2/2$.

Solve: The area of a sphere with a 100 ft radius is $A = 4\pi r^2 = 4\pi[(100\text{ ft})(\text{m}/(3.26\text{ ft}))]^2 = 1.17\times10^4\text{ m}^2$
Combine these expressions for the intensity and solve for the amplitude of the electric field.

$$E_0 = \left(\frac{2P}{c\varepsilon_0 A}\right)^{1/2} = \left(\frac{2(4.0\times10^{-3}\text{ W})}{(3.0\times10^8\text{ m/s})(8.85\times10^{-12}\text{ C}^2/(\text{N}\cdot\text{m}^2)(1.17\times10^4\text{ m}^2)}\right)^{1/2} = 0.016\text{ V/m}$$

Assess: Since the power output of the phone signal is small, we expect the amplitude of the oscillating electric field to be small.

P25.71. Prepare: Use Malus's law for the polarized light.

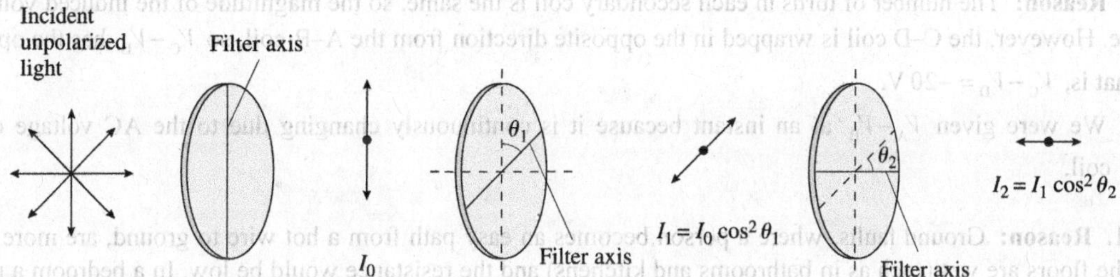

Solve: After passing through the first filter, the light is is polarized vertically and has intensity I_0. Now it passes through the second, inserted filter that is oriented at an angle $\theta_1 = 45°$ from the vertical. The light emerging from this filter will thus be polarized at $\theta_1 = 45°$ from the vertical and have intensity

$$I_1 = I_0 \cos^2 45° = \frac{I_0}{2}$$

Now the light passes through the last filter. You can see that the axis of this filter is at $45°$ with respect to the axis of polarization of the incoming light, so the intensity of the light after passing through this filter is

$$I_2 = I_1 \cos^2 45° = \frac{I_0}{2}\cos^2 45° = \frac{I_0}{4}$$

Assess: For this situation, the intensity is reduced by a factor of two each time it encounters a polarizing filter.

P25.73. Prepare: We will use the photon model of light, that is, Equation 25.22.
Solve: The energy of a 1000 kHz photon is $E_{photon} = hf = (6.63\times10^{-34}\text{ J}\cdot\text{s})(1000\times10^3\text{ Hz}) = 6.63\times10^{-28}\text{ J}$.

The energy transmitted each second is 20×10^3 J. The number of photons transmitted each second is 20×10^3 J/(6,63$\times$ 10^{-28} J) $= 3.0\times10^{31}$.

Assess: This number is large because of the significant power emitted by the antenna.

26

AC ELECTRICITY

Q26.1. Reason: In an AC circuit the average power dissipated is given by Equation 26.9: $P_R = (V_{rms})^2/R$.
The voltage given, 12 V AC, is the rms voltage. Since it is the same for both resistors then they dissipate the same power.
Assess: The frequency doesn't appear in the equation, so it doesn't affect the outcome.

Q26.5. Reason: We can use a transformer to step up the current. Knowing that $(I_2)_{rms}/(I_1)_{rms} = N_1/N_2$, we see that $(I_2)_{rms}$ can be 10 times greater than $(I_2)_{rms}$ if N_1 is 10 times greater than N_2.
Assess: A transformer with fewer turns at the secondary gives a greater current and a smaller voltage.

Q26.7. Reason: The number of turns in each secondary coil is the same, so the magnitude of the induced voltage is the same. However, the C–D coil is wrapped in the opposite direction from the A–B coil, so $V_C - V_D$ has the opposite sign. That is, $V_C - V_D = -20$ V.
Assess: We were given $V_A - V_B$ at an instant because it is continuously changing due to the AC voltage on the primary coil.

Q26.11. Reason: Ground faults, where a person becomes an easy path from a hot wire to ground, are more likely where the floors are wet (such as in bathrooms and kitchens) and the resistance would be low. In a bedroom a person isn't as likely to become a ground fault.
Assess: Look for GFI circuits in your kitchen and bathroom.

Q26.13. Reason: Knowing that the peak capacitor current is related to the given quantities by $I_C = 2\pi f C V_C$ and that $V_C = \mathcal{E}_0$, we see that the peak current will double when $\mathcal{E}_0$, C, or f are doubled. The peak current will increase by a factor of four if two of these quantities are doubled and by a factor of eight if all three are simultaneously doubled.
(a) The peak current doubles if the peak emf doubles. **(b)** The peak current doubles if the capacitance doubles.
(c) The peak current doubles if the frequency doubles.
Assess: Since the peak capacitor current is proportional to each of these quantities, it will double when they double.

Q26.17. Reason: Equation 26.20 gives the voltage across an inductor with a changing current: $v_L = L\Delta i_L/\Delta t$.
(a) For each inductor we are given v_L and L, but this only allows us to solve for $\Delta i_L/\Delta t$, not I_L itself. So we do not know which inductor has the largest current.
(b) Yes, we can tell in which inductor the current is changing most rapidly: $\Delta i_L/\Delta t = v_L/L$.
For the 2 H inductor, $\Delta i_L/\Delta t = v_L/L = 2$ V$/(2$ H$) = 1$ A/s.
For the 1 H inductor, $\Delta i_L/\Delta t = v_L/L = 4$ V$/(1$ H$) = 4$ A/s.

So the 1 H inductor has the current changing most rapidly.

Assess: Smaller inductances allow the current to change more easily.

Q26.21. Reason: The resonance frequency of an *RLC* circuit is given in Equation 26.26: $f_0 = 1/(2\pi\sqrt{LC})$.

(a) R is not in the equation, so doubling R does not change the resonance frequency; it is still 1000 Hz.

(b)

$$f_0 = \frac{1}{2\pi\sqrt{L'C}} = \frac{1}{2\pi\sqrt{2LC}} = \frac{1}{\sqrt{2}}f_0 = \frac{1}{\sqrt{2}}1000\,\text{Hz} = 710\,\text{Hz}$$

(c)

$$f_0 = \frac{1}{2\pi\sqrt{LC'}} = \frac{1}{2\pi\sqrt{L2C}} = \frac{1}{\sqrt{2}}f_0 = \frac{1}{\sqrt{2}}1000\,\text{Hz} = 710\,\text{Hz}$$

(d) $\mathcal{E}$ is not in the equation, so doubling $\mathcal{E}$ does not change the resonance frequency; it is still 1000 Hz.

(e) f is not in the equation, so doubling f does not change the resonance frequency; it is still 1000 Hz.

Assess: In addition to seeing what variables appear in the equation, we also ought to think physically about what quantities affect the resonance frequency. Even the driving frequency (which is the frequency at which the oscillator will oscillate) doesn't affect the *resonance* frequency, as the resonance frequency is only a function of the L and C of the circuit (the characteristics of the circuit itself, not how it is used or the emf driving it). Driving it at a different frequency just means the circuit is not at resonance.

Changing R doesn't affect the resonance frequency, but it will affect the current. Large R means small I.

Q26.23. Reason: The voltage of the secondary is given in Equation 26.13.

$$(V_2)_{\text{rms}} = \frac{N_2}{N_1}(V_1)_{\text{rms}} = \frac{100}{1000}120\,\text{V} = 12\,\text{V}$$

The current in the secondary is given in Equation 26.14.

$$(I_2)_{\text{rms}} = \frac{N_1}{N_2}(I_1)_{\text{rms}} = \frac{1000}{100}0.050\,\text{A} = 0.50\,\text{A}$$

The correct choice is D.

Assess: This is a step-down transformer, which reduces the voltage but increases the current.

Q26.27. Reason: The current is inversely proportional to $X_C = \frac{1}{2\pi fC}$ so increasing C increases the current and the brightness. The correct choice is A.

Assess: The effect is opposite for a variable inductor.

Q26.29. Reason: As the textbook says, at resonance the capacitor and inductor voltages have equal magnitudes but are exactly out of phase with each other, so they cancel when we add the voltages around the loop in Kirchhoff's loop law. So the peak voltage across the entire circuit is 8.0 V.

The correct choice is C.

Assess: We see that the resonance condition is met since $V_L = V_C$.

Problems

P26.1. Prepare: Please refer to Figure 26.3 for an AC resistor circuit.

Solve: (a) For a circuit with a single resistor, the peak current is

$$I_R = \frac{\mathcal{E}_0}{R} = \frac{10\,\text{V}}{200\,\Omega} = 0.050\,\text{A} = 50\,\text{mA}$$

(b) The peak current is the same as in part **(a)** because the current is independent of frequency.

Assess: Ohm's law applies to the peak current.

P26.3. Prepare: We have an AC one-resistor circuit.
Solve: From Equation 26.9,

$$R = \frac{V_{rms}^2}{P_{avg}} = \frac{(10.0\ \text{V})^2}{2.0\ \text{W}} = 50.0\ \Omega$$

Using the equation again,

$$V_{rms} = \sqrt{P_{avg} R} = \sqrt{(10.0\ \text{W})(50.0\ \Omega)} = 22.4\ \text{V}$$

Assess: As long as we work with rms voltages and currents, all the expressions for DC power carry over to AC power.

P26.7. Prepare: Equation 26.9 relates the rms current, the rms voltage (75 kV), and the average power (40 MW): $P_R = I_{rms} V_{rms}$.
Solve: Solve for I_{rms}.

$$I_{rms} = \frac{P_R}{V_{rms}} = \frac{40\ \text{MW}}{75\ \text{kV}} = 530\ \text{A}$$

Assess: This is a large current indeed, and it would require large transmission lines; we could make the current smaller by sending the power down the transmission lines at an even higher voltage.

P26.11. Prepare: We have been given the rms voltage and current for the secondary (output) of the transformer (power pack). We have also been given the rms voltage at the primary and we want to determine the rms current at the primary. The key to this problem is the knowledge that in the development of the transformer theory, we assumed the transformer was 100 percent efficient so the power at the secondary is the same as the power at the primary.
Solve: The power at the primary is equal to the power at the secondary: $P_1 = P_2$.

The power is related to the rms voltage and current by $P = V_{rms} I_{rms}$, which allows us to write $I_{1\,rms} V_{1\,rms} = I_{2\,rms} V_{2\,rms}$.

Solving for I_{1rms}, obtain

$$I_{1\,rms} = I_{2\,rms} V_{2\,rms} / V_{1\,rms} = (0.40\ \text{A})(5.2\ \text{V})/(120\ \text{V}) = 0.017\ \text{A}$$

Assess: Knowing that the power at the primary and secondary is the same, for the values given we should expect a small primary current.

P26.13. Prepare: According to Faraday's Law, the instantaneous voltage v_1 across the N_1 turns of the primary coil is $v_1 = N_1(\Delta \Phi_m / \Delta t)$ and the instantaneous voltage v_2 across the N_2 turns of the primary coil is $v_2 = N_2(\Delta \Phi_m / \Delta t)$. Transformer theory was developed based on the idea that the time rate of change of the magnetic flux is the same at the primary and secondary. This information may be used to determine the number of turns on the secondary coil. In the process of developing transformer theory we also assumed that the transformer was 100 percent efficient so the power at the primary was equal to the power at the secondary. Knowing that the power is related to the rms voltage and current by $P = V_{rms} I_{rms}$, we can determine the rms current at the primary.
Solve: (a) Knowing that the rate of change of the magnetic flux is the same at the primary and secondary we may write

$$\frac{\Delta \Phi_m}{\Delta t} = \frac{v_1}{N_1} = \frac{v_2}{N_2} \quad \text{or} \quad \frac{v_1}{v_2} = \frac{N_1}{N_2}$$

In this expression, v_1 and v_2 are instantaneous voltages, but since this expression is also valid at the instant the voltage is at its peak value we may write

$$\frac{V_1}{V_2} = \frac{N_1}{N_2} \quad \text{or} \quad \frac{V_{1\,rms}}{V_{2\,rms}} = \frac{N_1}{N_2}$$

which may be solved for N_2 to obtain

$$N_2 = N_1(V_{2\,rms} / V_{1\,rms}) = 400(4.5\ \text{V}/(120\ \text{V})) = 15$$

(b) The power at the primary is equal to the power at the secondary: $P_1 = P_2$.

The power is related to the rms voltage and current by $P = V_{rms} I_{rms}$, which allows us to write $I_{1\,rms} V_{1\,rms} = I_{2\,rms} V_{2\,rms}$. Solving for $I_{1\,rms}$, obtain

$$I_{1\,rms} = I_{2\,rms} V_{2\,rms} / V_{1\,rms} = (0.60\ \text{A})(4.5\ \text{V}) / (120\ \text{V}) = 0.0225\ \text{A}$$

which gives a peak current at the primary of

$$I_{1P} = I_{1\,rms} \sqrt{2} = (0.0225\ \text{A})\sqrt{2} = 0.032\ \text{A} = 32\ \text{mA}$$

Assess: Since this is a step-down transformer, we expect fewer turns at the secondary than at the primary and a smaller current at the primary than at the secondary.

P26.15. Prepare: First find the current in the wires: $I = P/\Delta V = 250\ \text{kW}/7.2\ \text{kV} = 34.7\ \text{A}$

Solve: (a) The power dissipated in the wires is now

$$P = I^2 R = (34.7\ \text{A})^2 (15\ \Omega) = 18\ \text{kW}$$

Now re-do the problem with the new voltage: $I = P/\Delta V = 250\ \text{kW}/30\ \text{kV} = 8.33\ \text{A}$

$$P = I^2 R = (8.33\ \text{A})^2 (15\ \Omega) = 1.0\ \text{kW}$$

Assess: We see there is less power lost to thermal dissipation at high voltages.

P26.19. Prepare: In regular household circuits these devices are all connected in parallel because they all want to see 120 V. We'll use $P = I\Delta V$ for each one to see how much current it will draw, and then add up the currents and compare to 15 A.

Solve: Blender: $I = P/\Delta V = 330\ \text{W}/120\ \text{V} = 2.8\ \text{A}$

Coffee pot: $I = P/\Delta V = 1000\ \text{W}/120\ \text{V} = 8.3\ \text{A}$

Coffee grinder: $I = P/\Delta V = 150\ \text{W}/120\ \text{V} = 1.3\ \text{A}$

Microwave oven: $I = P/\Delta V = 750\ \text{W}/120\ \text{V} = 6.3\ \text{A}$

The sum of these currents is $2.8\ \text{A} + 8.3\ \text{A} + 1.3\ \text{A} + 6.3\ \text{A} = 18.7\ \text{A}$, which is enough to trip the 15 A breaker.

Assess: The given values are realistic, so one should be careful to not turn everything on at once to avoid tripping the breaker.

P26.21. Prepare: We will first calculate the power dissipated in 1 year in kWh/year and then multiply it with the billing rate that is given as \$0.10/kWh.

Solve: The power dissipated by the night light in 1 year is

$$P = 60\ \text{watt} \times \frac{12\ \text{h}}{\text{d}} \times \frac{365\ \text{d}}{1\ \text{y}} = 263\ \text{kWh/year}$$

The cost of the night light in one year is $(\$0.10/\text{kWh})(263\ \text{kWh/year}) = \$26.30$.

Assess: An average cost of over \$2 per month for the nightlight is reasonable.

P26.27. Prepare: The total resistance is the resistance of the skin plus the resistance of two legs ($290\ \Omega$ each, from Figure 26.13). $R_{tot} = 1.0\ \text{k}\Omega + 2(290\ \Omega) = 1580\ \Omega$.

Solve:

$$I = \frac{\Delta V}{R} = \frac{300\ \text{V}}{1580\ \Omega} = 0.19\ \text{A} = 190\ \text{mA}$$

Assess: The large contact area and wet feet could make the skin resistance this low. This would be a dangerous DC current which would temporarily paralyze the respiratory muscles if the current went through the chest.

P26.29. Prepare: The total resistance is the resistance of the skin plus the resistance of two arms ($310\ \Omega$ each, from Figure 26.13) and his torso ($30\ \Omega$). $R_{tot} = 2(100\ \Omega) + 2(310\ \Omega) + 30\ \Omega = 850\ \Omega$.

Solve:

$$I = \frac{\Delta V}{R} = \frac{50\ \text{V}}{850\ \Omega} = 59\ \text{mA}$$

According to Table 26.1 he will certainly feel this DC current.

Assess: If the pulse were prolonged it would be dangerous and could cause paralysis of the respiratory muscles.

P26.31. Prepare: The total resistance is the resistance of the skin plus the resistance of two arms ($310\,\Omega$ each, from Figure 26.13). $R_{\text{tot}} = 2200\,\Omega + 2(310\,\Omega) = 2820\,\Omega$. The potential difference between the hands is computed as in the previous problem.

$$\Delta V = IR = (200\text{ A})(7.0\,\Omega)\left(\frac{0.15\text{ m}}{100\text{ km}}\right) = 2.1\text{ mV}$$

Solve:

$$I = \frac{\Delta V}{R} = \frac{2.1\text{ mV}}{2{,}820\,\Omega} = 7.4 \times 10^{-7}\text{ A}$$

This is undetectable, so yes, it is safe to hang from such a wire.

Assess: However, don't hook a leg on another wire which might be at a much different potential; that could easily be lethal.

P26.33. Reason: Knowing the peak current and voltage we can determine the capacitive reactance. Knowing the capacitive reactance and the capacitance, we can determine the oscillation frequency.

Solution: The capacitive reactance may be determined by $X_C = V_C/I_C$ and then the oscillation frequency may be determined by

$$f = \frac{1}{2\pi C X_C} = \frac{1}{2\pi C(V_C/I_C)} = \frac{I_C}{2\pi C V_C} = \frac{0.20\text{ A}}{2\pi(20\times10^{-6}\text{ F})(6.0\text{ V})} = 270\text{ Hz}$$

Assess: We obtained the correct units and this is not an unusually large frequency.

P26.37. Prepare: Figure 26.19 shows a simple AC capacitor circuit. We will use Equation 26.19 to find capacitive reactance.

Solve: The peak current through the capacitor is

$$I_C = \frac{V_C}{X_C} \Rightarrow X_C = \frac{V_C}{I_C} = \frac{1.0\text{ V}}{8.0\times10^{-3}\text{ A}} = 125\,\Omega = 130\,\Omega$$

Assess: Using reactance is just like using resistance in Ohm's law. This, however, does not apply to instantaneous values of current and voltage.

P26.39. Prepare: Figure 26.20 shows an AC one-inductor circuit.

Solve: (a) The peak current through the inductor is

$$I_L = \frac{V_L}{X_L} = \frac{V_L}{2\pi f L} = \frac{10.0\text{ V}}{2\pi(100\text{ Hz})(20\times10^{-3}\text{ H})} = 0.80\text{ A}$$

(b) At a frequency of 100 kHz instead of 100 Hz as in part **(a)**, the reactance will increase by a factor of 1000 and thus the current will decrease by a factor of 1000. Thus, $I_L = 0.80$ mA.

Assess: The current is inversely proportional to the reactance.

P26.45. Prepare: A capacitor in combination with an inductor will have a resonant frequency determined by the inductance and capacitance. Changing the inductance in combination with a capacitance will change the resonant frequency of the *LC* circuit. We will use Equation 26.25.

Solve: We know the frequency and the capacitance so we can find the inductance. We have

$$f = \frac{1}{2\pi}\sqrt{\frac{1}{LC}} \Rightarrow L = \frac{1}{(2\pi f)^2 C} = \frac{1}{(2\pi\times100\times10^{6}\text{ Hz})^2(10\times10^{-12}\text{ F})} = 2.53\times10^{-7}\text{ H} = 0.25\ \mu\text{H}$$

Assess: All devices that send or receive electromagnetic waves use an *LC* circuit like this to establish the correct frequency.

P26.47. Prepare: We will equate the capacitive and inductive reactances and thus find the resonant frequency. This frequency can then be used to calculate reactances using Equations 26.18 and 26.24.

Solve: When a capacitive reactance X_C and an inductive reactance X_L become equal,

$$2\pi fL = \frac{1}{2\pi fC} \Rightarrow f = \frac{1}{2\pi}\sqrt{\frac{1}{LC}} \Rightarrow f = \frac{1}{2\pi}\frac{1}{\sqrt{(1.0\times10^{-6}\,\text{H})(1.0\times10^{-6}\,\text{F})}} = 160\,\text{kHz}$$

The reactance at this frequency is

$$X_C = X_L = \omega L = 2\pi(160\,\text{kHz})(1.0\times10^{-6}\,\text{H}) = 1.0\,\Omega$$

Assess: All devices that send or receive electromagnetic waves use an LC circuit like this to establish the correct frequency.

P26.49. Prepare: At the resonance frequency, the current in the series RLC circuit is a maximum. The resistor does not affect the resonance frequency, so we can use Equation 26.26.
Solve: From Equation 26.26,

$$f = \frac{1}{2\pi}\sqrt{\frac{1}{LC}} \Rightarrow L = \frac{1}{4\pi^2 f^2 C} = \frac{1}{4\pi^2(1000\,\text{Hz})^2(2.5\times10^{-6}\,\text{F})} = 10\,\text{mH}$$

Assess: The inductance obtained is reasonable.

P26.53. Prepare: The transformer equation says that

$$\frac{(V_2)_{\text{rms}}}{(V_1)_{\text{rms}}} = \frac{N_2}{N_1}$$

where $(V_2)_{\text{rms}} = 230\,\text{V}$ and $(V_1)_{\text{rms}} = 120\,\text{V}$.
Solve: The kettle will still draw 13 A from the secondary coil, so the primary coil must have a current of

$$I_1 = I_2\frac{N_2}{N_1} = I_2\frac{(V_2)_{\text{rms}}}{(V_1)_{\text{rms}}} = (13\,\text{A})\frac{230\,\text{V}}{120\,\text{V}} = 25\,\text{A}$$

Assess: This current will certainly trip a 15 A breaker.

P26.57. Prepare: Given the expression for the instantaneous voltage across the inductor, we can determine the voltage across the resistor at a specified time. Knowing the inductance and how to obtain the frequency from the expression for the instantaneous voltage, we can determine the inductive reactance.
Solve:
(a) The instantaneous voltage at the specified time is determined by

$$v_L = V_L\cos 2\pi ft = (25\,\text{V})\cos[(60\,\text{Hz})(0.10\,\text{s})] = 24\,\text{V}$$

(b) The inductive reactance is

$$X_L = 2\pi fL = (60\,\text{Hz})(75\times10^{-6}\,\text{H}) = 4.5\,\text{m}\Omega$$

(c)

$$I_L = \frac{V_L}{X_L} = \frac{25\,\text{V}}{4.5\times10^{-3}\,\Omega} = 5600\,\text{A}$$

Assess: The peak current is quite large, but all the units work out.

P26.59. Prepare: This is an RLC circuit.
Solve: (a) The resonance frequency of the circuit is

$$f = \frac{1}{2\pi}\frac{1}{\sqrt{LC}} = \frac{1}{2\pi}\frac{1}{\sqrt{(10\times10^{-3}\,\text{H})(10\times10^{-6}\,\text{F})}} = 500\,\text{Hz}$$

(b) At resonance, $X_L = X_C$. So, the peak current is

$$I = \frac{\mathcal{E}_0}{R} = \frac{10\,\text{V}}{10\,\Omega} = 1.0\,\text{A}$$

Assess: At resonance, the circuit is resistive. We can use Ohm's law with the peak voltage and the value of R to find the peak current. The current obtained is reasonable.

PART VI

ELECTRICITY AND MAGNETISM

PptVI.7. Reason: Use Newton's second law.

$$a = \frac{F}{m} = \frac{qE}{m} = \frac{q\frac{\Delta V}{d}}{m} = \frac{(1.6\times10^{-19}\,\text{C})(70\,\text{kV}/1.0\,\text{cm})}{9.11\times10^{-31}\,\text{kg}} = 1.2\times10^{18}\ \text{m/s}^2$$

The correct choice is A.

Assess: This is a surprisingly large acceleration, but so were the other choices, so we trust the math.

PptVI.9. Reason: We are given $I = 10\,\text{mA}$ and $\Delta V = 70\,\text{kV}$.

$$P = I\Delta V = (10\,\text{mA})(70\,\text{kV}) = 700\ \text{W}$$

The correct choice is B.

Assess: This is a decent amount of power, but it isn't on for very long.

PptVI.11. Reason: Assume the maximum energy of the x-ray photon is the kinetic energy of the electron just before it strikes the target electrode.

$$E = \frac{1}{2}mv^2 = \frac{1}{2}m(2a\Delta x) = ma\Delta x = (9.11\times10^{-31}\,\text{kg})(1.2\times10^{18}\,\text{m/s}^2)(0.010\,\text{m}) = 1.1\times10^{-14}\ \text{J}$$

The correct choice is C.

Assess: For a photon this is fairly energetic (68 keV), as we expect for an x-ray.

PptVI.15. Reason: The energy stored in a capacitor is $\frac{1}{2}C(\Delta V)^2$. Halving the voltage cuts the energy stored to 1/4 the original value. The correct choice is D.

Assess: This is a challenge for a capacitor-powered car.

PptVI.17. Reason: The induced current in the secondary will be in such a direction as to oppose the increase of the field from the primary, so the induced current will create a field to the left. The correct choice is B.

Assess: If the right-pointing field from the primary were decreasing, the induced current would be in the other direction, such as to oppose the decrease.

PptVI.21. Reason: Each side of the wire loop has a mass of 50 g. To hold the loop steady we must have $\Sigma\tau = 0$. Compute the torques around the pivot axis. The downward torque due to the weight of the two sides of the loop perpendicular to the axis plus the torque due to the weight of the side parallel to the axis is

$$\tau_{\text{grav}} = 2(50\,\text{g})(9.8\,\text{m/s}^2)(5.0\,\text{cm}) + (50\,\text{g})(9.8\,\text{m/s}^2)(10\,\text{cm}) = 0.098\ \text{N}\cdot\text{m}$$

The upward torque due to the current in the magnetic field must counterbalance that torque. No torque is generated in the two sides perpendicular to the axis because $\sin\theta = 0$, so all the torque must be generated on the side parallel to the axis. The current will need to be clockwise (looking from above) to produce the upward torque, and in the parallel side $\sin\theta = 1$.

$$\tau_{mag} = r_\perp F_{mag} = L(ILB) = IL^2 B \quad \Rightarrow \quad I = \frac{\tau}{L^2 B} = \frac{0.098\ \text{N} \cdot \text{m}}{(0.10\ \text{m})^2 (0.50\ \text{T})} = 19.6\ \text{A} \approx 20\ \text{A}$$

Assess: It seems reasonable that a current of this magnitude could hold up a 200 g loop of wire.

27

RELATIVITY

Q27.1. Reason: If we assume that the airplane's velocity is constant (so you are in an inertial reference frame), then there is no experiment you can do to prove that you are moving and people on the ground (in another reference frame) are at rest.

Assess: We have also assumed that the people on the ground are in an inertial reference frame. In a very picky sense, neither the plane's nor the ground's reference frame is inertial because both of them are accelerating around the sun, for example. But in this chapter we will neglect the centripetal acceleration of the earth and assume the reference frames are inertial.

Q27.3. Reason: We'll assume that relativity isn't important for the speeds in this question so we will use the Galilean transformation Equations 27.1: $u' = u - v$ and $u = u' + v$.

(a) We'll call the reference frame of the page S, and that of ball 1 S'. Then the velocity of the ball's reference frame relative to ours is $v = 6$ m/s, $u_1 = 6$ m/s, and $u_2 = -3$ m/s.

$$u'_1 = u_1 - v = 6 \text{ m/s} - 6 \text{ m/s} = 0 \text{ m/s}$$

$$u'_2 = u_2 - v = -3 \text{ m/s} - 6 \text{ m/s} = -9 \text{ m/s} \qquad \text{speed: 9 m/s, direction: to the left}$$

(b) We'll call the reference frame of the page S, and that of ball 2 S'. Then the velocity of the ball's reference frame relative to ours is $v = -3$ m/s, $u_1 = 6$ m/s, and $u_2 = -3$ m/s.

$$u'_1 = u_1 - v = 6 \text{ m/s} - (-3 \text{ m/s}) = 9 \text{ m/s} \qquad \text{speed: 9 m/s, direction: to the right}$$

$$u'_2 = u_2 - v = -3 \text{ m/s} - (-3 \text{ m/s}) = 0 \text{ m/s}$$

Assess: The answers here are simple and intuitive; you didn't need to be elaborate about the equations to see the answers in this case. But it is important to discipline yourself to do so in order to identify the various reference frames and variables correctly—especially when we can't use the simple Galilean transformation equations.

Q27.7. Reason: Event 1 occurs after event 2 because the flash of light from event 2 had to travel twice as far as the flash of light from event 1 and will take twice as long to travel that longer distance.

Assess: Since the two firecrackers are in the same direction away from you the light from event 2 had to pass by the location of event 1 before it could continue on to you.

Q27.11. Reason: (a) Since Peggy is in the same reference frame as the firecrackers and they are equidistant from her and she saw the two flashes at the same time then yes, the explosions were simultaneous in Peggy's reference frame.

(b) In Ryan's frame the two explosions are not simultaneous. The left one occurred first because it had a farther distance to travel to reach Peggy, since Peggy was moving toward the right and away from the left.

Assess: Simultaneity is one of our cherished concepts we must give up in relativity. Two events that are simultaneous in one inertial reference frame will not be simultaneous in other inertial reference frames.

Q27.13. Reason: To measure the length of a meter stick you would note the time on your clock when the leading edge of the stick is exactly abreast of you and then do the same for the trailing edge as it passes you. The length of the stick is $L = \Delta x = v\Delta t$, where $v = 0.5c$.

Assess: Your result will be less than a meter.

Q27.15. Reason: (a) You measure the proper length l because you measure it in a reference frame in which the object (the land between LA and NY) is at rest.

(b) Your friend is in a reference frame that is moving with respect to yours, and so she will measure a distance L' which is shorter than l. Equation 27.12 shows that $L' < l$: $L' = l\sqrt{1 - \beta^2} < l$.

Assess: Everyone in reference frames other than the one in which the object is at rest will measure the length of the object to be less than l.

Q27.19. Reason: The total energy of a particle is given in Equation 27.19: $E = \gamma mc^2$.

But $\gamma \geq 1$ so the energy can never be less than than mc^2.

Assess: The rest energy is given by Equation 27.21: $E_0 = mc^2$. If the particle has any kinetic energy (i.e., is moving) then the energy will be greater than the rest energy.

Q27.21. Reason: Review the section on the twins paradox and Equations 27.7 and 27.8. We'll assume the turnaround time is very short so that nearly all of Lee's trip is at the speed of $0.866c$ (although it is important to note that Lee does not stay in one single inertial reference frame for the whole trip). We can treat this as a time dilation problem. Since we want to know how much Lee ages, we use Lee's clock. Therefore Lee measures (and ages) the proper time $\Delta\tau$, while Leigh measures (and ages) a time interval $\Delta t_{\text{Leigh}} = (11 \text{ years old}) - (1 \text{ year old})$ $= 10$ years.

$$\Delta\tau = \sqrt{1 - \beta^2}\,\Delta t_{\text{Leigh}} = \sqrt{1 - 0.866^2}\,(10 \text{ years}) = 5 \text{ years}$$

Lee aged 5 years and so is now 6 years old.

The correct choice is B.

Assess: We could also have guessed the answer by elimination of other choices. Surely Lee will age some, so choice A is not correct. And because this is a twins-paradox problem we know that Lee won't age the same amount as Leigh, so C is not correct. But we also know the traveling twin ages less than the stay-at-home twin, so D is not correct.

Q27.23. Reason: To measure the proper time the astronaut with her clock needs to have both events occur at the same position in her reference frame. Call that position the origin (the location of the astronaut). She needs to travel 150 m in 1 μs to be present at the second flash. The speed required is

$$v = \frac{150 \text{ m}}{1\,\mu s} = 1.5 \times 10^8 \text{ m/s} = 0.5c$$

The correct choice is B.

Assess: The measurement of the time interval by the astronaut would be a smaller value than measuring it in any other inertial reference frame.

Problems

P27.1. Prepare: S is the ground's frame of reference and S′ is the sprinter's frame of reference. Frame S′ moves relative to frame S with speed v.

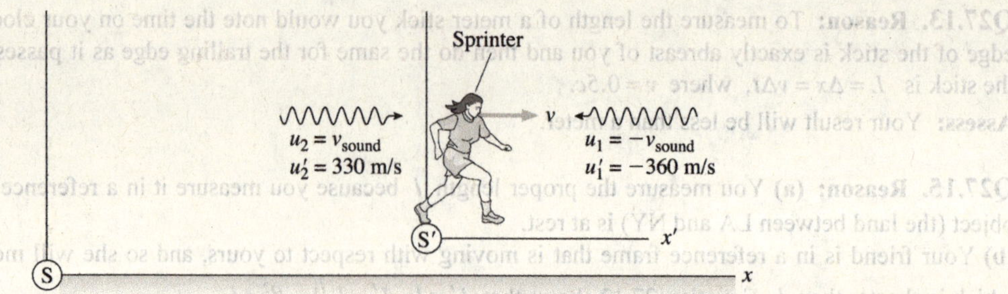

Solve: The speed of a sound wave is measured relative to its medium. The medium is still air on the ground, which is our frame S. The sprinter travels to the right with reference frame S′ at velocity v. Using the Galilean transformations of velocity,

$$u'_1 = -355 \text{ m/s} = u_1 - v = -v_{sound} - v \qquad u'_2 = 335 \text{ m/s} = u_2 - v = v_{sound} - v$$

Adding the two previous equations,

$$-20 \text{ m/s} = -2v \Rightarrow v = 10 \text{ m/s}$$

From the first equation,

$$-335 \text{ m/s} = -v_{sound} - (10 \text{ m/s}) \Rightarrow v_{sound} = 345 \text{ m/s}$$

Assess: Notice that the Galilean transformations use velocities and not speeds. It is for that reason $u'_1 = -355$ m/s. The speed of sound is what we expected. The speed of the sprinter is world-class, but possible.

P27.5. Prepare: Assume motion along the x-direction. Let the earth frame be S and a frame attached to the moving sidewalk be S′. Frame S′ moves relative to S with velocity V_x.

Solve: Let vx be your velocity in frame S and v'_x be your velocity in S′. In the first case, when the moving sidewalk is broken, $V_x = 0$ m/s and

$$v_{xW} = \frac{(x_1 - x_0)}{50 \text{ s}}$$

In the second case, when you stand on the moving sidewalk, $v'_x = 0$ m/s. Therefore, using $v_x = v'_x + V_x$, we get

$$v_{xS} = V_x = \frac{x_1 - x_0}{75 \text{ s}}$$

In the third case, when you walk while riding, $v'_x = v_{xW}$. Using $v_x = v'_x + V_x$, we get

$$\frac{x_1 - x_0}{t} = \frac{x_1 - x_0}{50 \text{ s}} + \frac{x_1 - x_0}{75 \text{ s}} \Rightarrow t = 30 \text{ s}$$

Assess: A time smaller than 50 s was expected.

P27.7. Prepare: Assume the spacecraft is an inertial reference frame.
Solve: Light travels at speed c in all inertial reference frames, regardless of how the reference frames are moving with respect to the light source. Relative to the spacecraft, the starlight is approaching at the speed of light $c = 3.0 \times 10^8$ m/s.

P27.11. Prepare: Bjorn and firecrackers 1 and 2 are in the same reference frame. Light from both firecrackers travels towards Bjorn at 300 m/μs. Bjorn is 600 m from the origin.

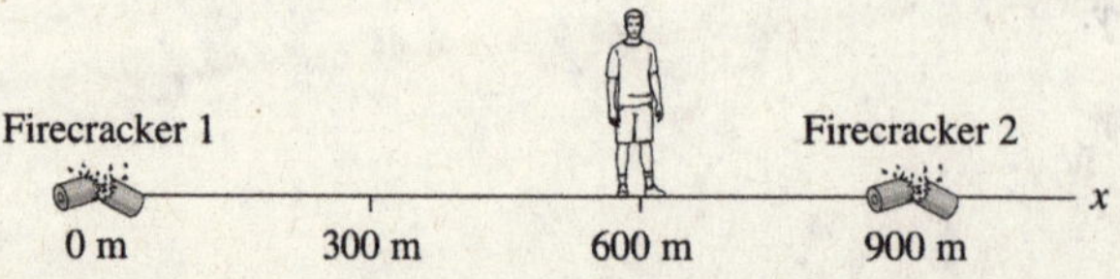

Solve: Light with a speed of 300 m/μs takes 2.0 μs to reach Bjorn. Since this flash reaches Bjorn at $t = 3.0$ μs, it left firecracker 1 at $t_1 = 1.0$ μs. The flash from firecracker 2 takes 1.0 μs to reach Bjorn. So, the light left firecracker 2 at $t_2 = 2.0$ μs.

Assess: Note that the two events are *not* simultaneous although Bjorn *sees* the events as occurring at the same time.

P27.13. Prepare: You and your assistant are in the same reference frame. Light from the two lightning bolts travels toward you and your assistant at 300 m/μs. You and your assistant have synchronized clocks.

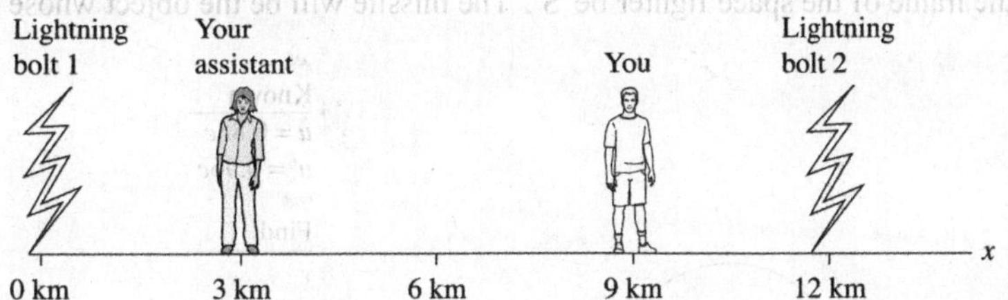

Solve: Bolt 1 is 9 km away, so it takes 30 μs for the light to reach you $(9000$ m $\div 300$ m/μs$)$. Bolt 2 is 3 km away from you, so it takes 10 μs to reach you. Since both flashes reach your eye at the same time, event 1 happened 20 μs before event 2. If event 1 happened at time $t_1 = 0$ then event 2 happened at time $t_2 = 20$ μs. For your assistant, it takes light from bolt 1 10 μs to reach her and light from bolt 2 30 μs to reach her. She sees the flash from bolt 1 at $t = 10$ μs and the flash from bolt 2 at $t = 50$ μs. That is, your assistant sees flash 2 40 μs after she sees flash 1.

P27.17. Prepare: The light travels at c in both reference frames.
Solve: **(a)** In the frame of the ship the light has to go the same distance in both directions, so it gets to both ends at the same time. The events are simultaneous in the frame of the ship. **(b)** From the earth's frame, the ship moves while the light is enroute, so the rear of the ship comes up to meet the light while the front of the ship moves farther away while the light is enroute. In the frame of the earth event 2 occurs before event 1, so they are not simultaneous.

P27.21. Prepare: The earth is frame S and the starship is frame S$'$. S$'$ moves relative to S with a speed v.
Solve: **(a)** The speed of the starship is

$$v = \frac{20 \text{ ly}}{25 \text{ years}} = \frac{(20 \text{ years})c}{25 \text{ years}} = 0.80c$$

(b) The astronauts measure the proper time while they are traveling. This is

$$\Delta\tau = \sqrt{1 - \frac{v^2}{c^2}}\,\Delta t = \sqrt{1 - (0.8)^2}\,(25 \text{ years}) = 15 \text{ years}$$

Because the explorers stay on the planet for one year, the time elapsed on their chronometer is 16 years.
Assess: The results are consistent with the discussion in Section 27.6.

P27.23. Prepare: S$'$ is the rocket's frame (Jill's frame) and S is the ground's frame (your frame). In the S$'$ frame, which moves with a velocity v relative to S, the length of the rocket is the proper length because it is at rest in this frame. So, $L' = l$.
Solve: You measure a length-contracted rocket

$$L = \sqrt{1 - \beta^2}\, l \Rightarrow 80 \text{ m} = \sqrt{1 - \beta^2}\,(100 \text{ m}) \Rightarrow \beta = 0.60$$

She did exceed the $0.5c$ speed limit.

P27.27. Prepare: Call the contraction $\Delta L = l - L = l - l\sqrt{1-\beta^2}$.
Solve: Use the binomial approximation.

$$\Delta L = l\left(1-\sqrt{1-\beta^2}\right) = l(1-(1-\tfrac{1}{2}\beta^2)) = l(\tfrac{1}{2}\beta^2) = (15.5 \text{ m})\left(\frac{1}{2}\left(\frac{2020 \text{ m/s}}{3.0\times10^8 \text{ m/s}}\right)^2\right) = 0.35 \text{ nm}$$

Assess: If the binomial approximation isn't used the answer is 0.40 nm.

P27.29. Prepare: This is a basic relativistic velocity addition problem so we can use Equation 27.15. Let the frame of Planet X be S and the frame of the space fighter be S'. The missile will be the object whose velocity is measured.

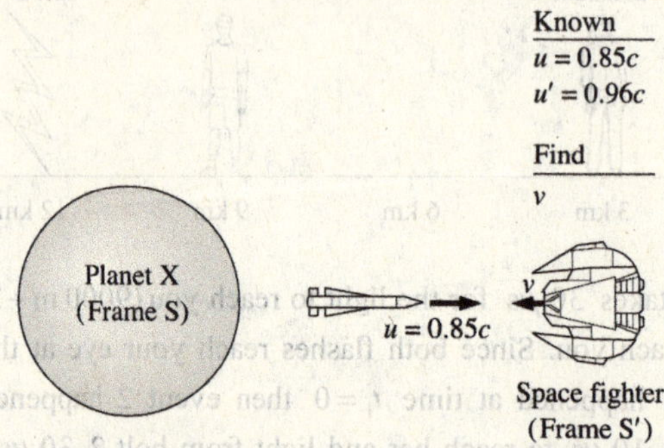

Solve: The velocity of the missile relative to Planet X (at rest in frame S) is $0.85c$ so $u = 0.85c$. Since the velocity of the missile relative to the space fighter is $0.96c$, we can write: $u' = 0.96c$. The velocity of the space fighter relative to Planet X is the velocity of frame S' relative to S, that is v. We can find v from Equation 27.15:

$$u' = \frac{u-v}{1-uv/c^2} \Rightarrow u'(1-uv/c^2) = u-v \Rightarrow v - u'uv/c^2 = u - u' \Rightarrow$$

$$v = \frac{u-u'}{1-u'u/c^2} = \frac{0.85c - 0.96c}{1-(0.96c)(0.85c)/c^2} = -0.60c$$

The speed of the space fighter relative to Planet X is $0.60c$.

Assess: Notice that when we solved Equation 27.15 for v, we ended up with an equation, namely $v = \frac{u-u'}{1-u'u/c^2}$, just like the one we started with, but with u' and v switched. This is a reasonable result since in the second part of Equation 27.15, the formula for u, we can see that u' and v are interchangeable.

P27.31. Prepare: We will use Equations 27.17 and 27.18.
Solve: (a) The relativistic momentum is

$$p = \frac{mu}{\sqrt{1-u^2/c^2}} = \frac{(1.67\times10^{-27} \text{ kg})(0.999)(3.0\times10^8 \text{ m/s})}{\sqrt{1-(0.999)^2}} = 1.12\times10^{-17} \text{ kg} \cdot \text{m/s}$$

(b) The ratio of the relativistic momentum and the Newtonian momentum is

$$\frac{p_{\text{relativistic}}}{p_{\text{classical}}} = \frac{mu}{\sqrt{1-u^2/c^2}}\frac{1}{mu} = \frac{1}{\sqrt{1-u^2/c^2}} = 22.4$$

Assess: We must use relativistic formulas when the speeds are comparable to the speed of light.

P27.33. Prepare: We will use Equations 27.17 and 27.18.
Solve: The Newtonian momentum is $p_{\text{Newton}} = mu$. We have

$$p = \frac{mu}{\sqrt{1-u^2/c^2}} = 2\,mu \Rightarrow 1-u^2/c^2 = \frac{1}{4} \Rightarrow u = \frac{\sqrt{3}}{2}c = 0.87c$$

Assess: As previously obtained, we expected the particle to be moving with very high speed.

P27.37. Prepare: The total energy of an object of mass m moving at speed u is $E = \gamma E_0$.
Solve: We have

$$E = (1.10)E_0 = \gamma E_0 \Rightarrow \gamma = 1.10 = \frac{1}{\sqrt{1 - u^2/c^2}} \Rightarrow u = 0.42c$$

P27.41. Prepare: The two events of interest are the creation and annihilation of the particle. These two events occur at the same place in a frame traveling with the particle, hence proper time is measured in the frame traveling with the particle. The time between these two events in the lab frame is $\Delta t = 2.3$ ps. Proper time is related to the lab time by $\Delta \tau = \Delta t \sqrt{1 - (v/c)^2}$.
Solve: The lifetime of the particle according to the particle is the proper time and is determined by the following:

$$\Delta \tau = \Delta t \sqrt{1 - (v/c)^2} = (2.3\,\text{ps})\sqrt{1 - (0.9c/c)^2} = 1.0\,\text{ps}$$

Assess: This answer is consistent with the knowledge that the moving clock runs slow.

P27.43. Prepare: We need to know how much the car is length contracted relative to David so we need to know the speed of the car relative to David. For that we can use the relativistic velocity transformation of Equation 27.15. Let David's frame of reference be S' and your frame of reference be S.

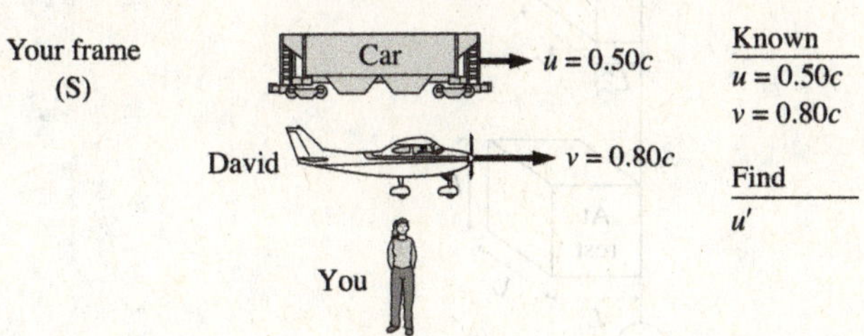

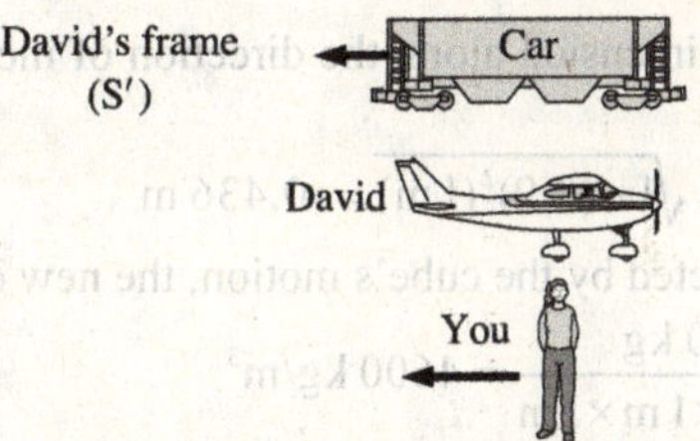

Solve: The velocity of S' relative to S is $v = 0.80c$ and the velocity of the car relative to you is $u = 0.50c$. The velocity of the car relative to David is as follows:

$$u' = \frac{u - v}{1 - uv/c^2} = \frac{0.50c - 0.80c}{1 - (0.50c)(0.80c)} = -0.50c$$

Since the car has the same speed relative to David that it has relative to you, $0.50c$, he observes the same length contraction that you observe. David measures the car to be 12 m.
Assess: As a check on our work, we can find the velocity of the car relative to you given the velocity of the car relative to Dave. (We hope to obtain $0.50c$.) Using the second part of Equation 27.15, we obtain the following:

$$u = \frac{u' + v}{1 + u'v/c^2} = \frac{-0.50c + 0.80c}{1 + (-0.50c)(0.80c)/c^2} = 0.50c$$

in agreement with the statement of the problem.

P27.47. Prepare: Even in relativity $\Delta x = v\Delta t$.

Solve: In the sun's frame it will take $\Delta t = \dfrac{5\times10^9 \text{ m}}{(1.7)(3\times10^8 \text{ m/s})} = 9.80$ s for the ships to pass each other. Use the time dilation equation for each spaceship. For the $0.8c$ ship, $\Delta t = \dfrac{9.80 \text{ s}}{\sqrt{1-(0.8)^2}} = 16.3$ s, and for the $0.9c$ ship,

$\Delta t = \dfrac{9.80 \text{ s}}{\sqrt{1-(0.9)^2}} = 22.5$ s. The difference is 6.2 s.

Assess: Relative to the sun the faster ship's clock runs slower than the slower ship's.

P27.49. Prepare: The length of an object is contracted when it is measured in any reference frame moving relative to the object. The contraction occurs only along the direction of motion of the object. The cube at rest has a density of 2000 kg/m^3. That is, a cube of $1\text{ m}\times1\text{ m}\times1\text{ m}$ dimensions has a mass of 2000 kg.

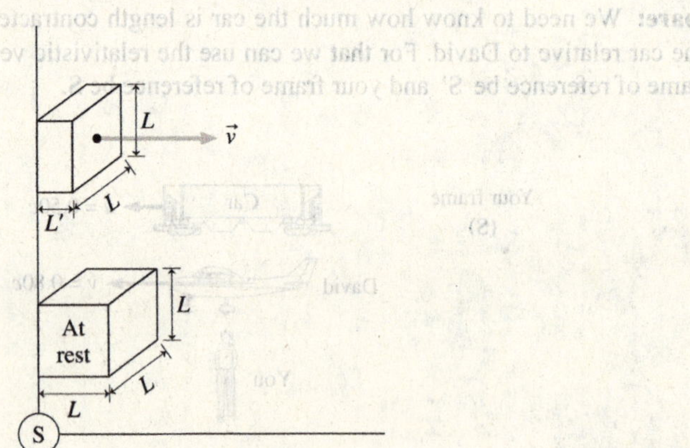

Solve: As the cube moves with a speed of $0.9c$, its dimension along the direction of motion is contracted according to Equation 27.12:

$$L' = \sqrt{1-(v/c)^2}\,L = \sqrt{1-(0.9)^2}(1\text{ m}) = 0.436 \text{ m}$$

Because the other dimensions of the cube are not affected by the cube's motion, the new density will be

$$\rho = \frac{2000 \text{ kg}}{0.436 \text{ m}\times1\text{ m}\times1\text{ m}} = 4600 \text{ kg/m}^3$$

Assess: The increased density is due to smaller volume because of length contraction along the direction of motion.

P27.53. Prepare: In problems of this type in which someone travels at high speed to some destination, we need to bear in mind that the length of the journey is subject to length contraction and, whereas the time elapsed as measured by someone on the spaceship is the proper time, for any other observer, the time taken for the trip is dilated.

Solve: **(a)** For a scientist on earth, the time taken is simply $\Delta t = l/v = (9.0 \text{ ly})/(0.90c) = 10$ y.

(b) The factor γ is $\dfrac{1}{\sqrt{1-v^2/c^2}} = \dfrac{1}{\sqrt{1-(0.9)^2}} = 2.29$. The scientist on the spaceship measures the proper time, as given by Equation 27.6.

$$\Delta\tau = \Delta t\sqrt{1-v^2/c^2} = \Delta t/\gamma = (10 \text{ yr})/(2.29) = 4.4 \text{ y}$$

(c) According to the scientist on the spaceship, the path length of the trip is length contracted to the following:

$$L = l\sqrt{1-v^2/c^2} = l/\gamma = (9.0 \text{ ly})/(2.29) = 3.9 \text{ ly}$$

(d) As in classical physics, in Einstein's relativity, two observers measure equal and opposite velocities for one another. Since the speed of the spaceship relative to the star is $0.90c$, the speed of the star relative to the spaceship is also $0.90c$. They see the star approach at $0.90c$.

Assess: As a check on our answer in part **(d)**, we could divide the distance to the star by the time it takes the star to reach the spaceship. (In the spaceship's reference frame, the star is doing the traveling.)

$$L/\Delta\tau = (3.93 \text{ ly})/(4.37 \text{ yr}) = 0.90c$$

This is as we obtained before.

P27.55. Prepare: Let S be the earth's reference frame and S$'$ be the rocket's reference frame. S$'$ travels at $0.5c$ relative to S.
Solve: **(a)** For the earthlings, the total distance traveled by the rocket is $2 \times 4.25 \text{ ly} = 8.50 \text{ ly}$. The time taken by the rocket for the round trip is

$$\frac{8.50 \text{ ly}}{0.500c} = \frac{8.50 \text{ ly}}{0.500 \text{ ly/y}} = 17.0 \text{ y}$$

(b) The time interval measured in the rocket's frame S$'$ is the proper time because it can be measured with a single clock at the same position. So,

$$\Delta t = \frac{\Delta\tau}{\sqrt{1-(v/c)^2}} \Rightarrow 17 \text{ y} = \frac{\Delta\tau}{\sqrt{1-(0.5)^2}} \Rightarrow \Delta\tau = 14.7 \text{ y}$$

The distance traveled by the rocket crew is length contracted to

$$L' = L\sqrt{1-(v/c)^2} = (8.50 \text{ ly})\sqrt{1-(0.5)^2} = 7.36 \text{ ly}$$

Note that the speed is still the same:

$$v = \frac{L'}{\Delta\tau} = \frac{7.36 \text{ ly}}{14.7 \text{ yr}} = 0.500c$$

(c) Both are correct in their own frame of reference.

P27.57. Prepare: Let S be the earth's reference frame and S$'$ be the reference frame of one rocket. S$'$ moves relative to S with $v = -0.75c$. The speed of the second rocket in the frame S is $u = +0.75c$.

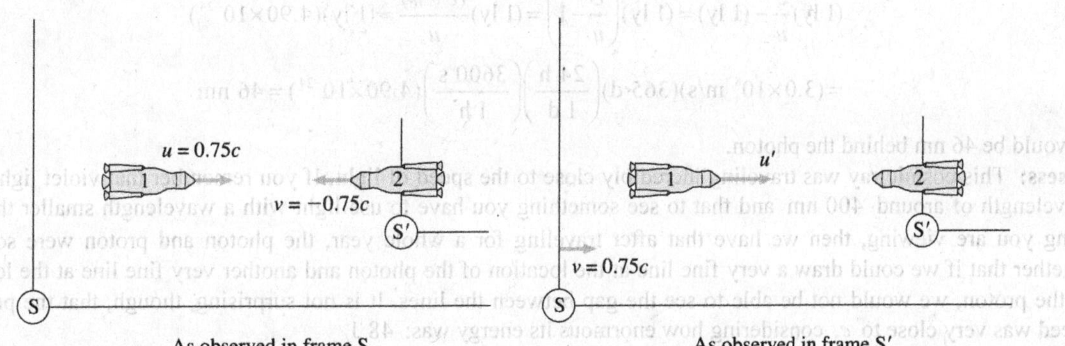

Solve: Using the Lorentz velocity transformation Equation 27.15,

$$u' = \frac{u-v}{1-uv/c^2} = \frac{0.75c-(-0.75c)}{1-(0.75c)(-0.75c)/c^2} = 0.96c$$

Assess: In Newtonian mechanics, the Galilean transformation of velocity will give $u' = 0.75c - (-0.75c) = 1.50c$. This is not permissible according to the theory of relativity.

P27.61. Prepare: Because the ship moves between pulses even in Karma's frame the pulses reach the ship at less than 1 second intervals. Think of the pulses as having a 1 s frequency: $f_{source} = 1.0$ Hz. We are given $v = 0.95c$.
Solve: The ship moves a distance $d = c/f_{source}$ between pulses, so in Karma's frame the pulses hit the ship at time intervals of

$$\Delta t = \frac{d}{c+v} = \frac{c}{(c+v)f_{source}} = \frac{1}{(1+\beta)f_{source}} = 0.513 \text{ s}$$

Now we have to shift this result into Trevor's reference frame by multiplying by $\sqrt{1-\beta^2}$. So in Trevor's frame the time between pulses is

$$(0.513 \text{ s})\sqrt{1-0.95^2}=0.16 \text{ s}$$

Assess: We expected the answer to be less than 1 s.

P27.63. Prepare: Since we have the proton's total energy, and we can look up its mass, we can find the speed of the proton using Equation 27.19.
Solve: (a) Solving Equation 27.19 for v gives:

$$E=\frac{mc^2}{\sqrt{1-u^2/c^2}}\Rightarrow\sqrt{1-u^2/c^2}=\frac{mc^2}{E}\Rightarrow u=\sqrt{1-\left(\frac{mc^2}{E}\right)^2}\,c$$

The quantity $\left(\dfrac{mc^2}{E}\right)^2$ is very small:

$$\left(\frac{mc^2}{E}\right)^2=\left(\frac{(1.67\times10^{-27}\text{ kg})(3.0\times10^8\text{ m/s})^2}{(3.0\times10^{20}\text{ eV})}\frac{1\text{ eV}}{1.60\times10^{-19}\text{ J}}\right)^2=9.80\times10^{-24}$$

So we will use the binomial approximation:

$$u=\sqrt{1-\left(\frac{mc^2}{E}\right)^2}\,c\approx\left(1-\frac{1}{2}\left(\frac{mc^2}{E}\right)^2\right)c=(1-4.90\times10^{-24})c$$

The proton's speed was given by $u=(1-4.9\times10^{-24})c$.

(b) The time it would take for the proton to travel 1 ly is $(1\text{ ly})/u$. In that time, the photon would have traveled a distance $(1\text{ ly})c/u$. The extra distance traveled by the photon is as follows:

$$(1\text{ ly})\frac{c}{u}-(1\text{ ly})=(1\text{ ly})\left(\frac{c}{u}-1\right)=(1\text{ ly})\frac{(c-u)}{u}\approx(1\text{ ly})(4.90\times10^{-24})$$

$$=(3.0\times10^8\text{ m/s})(365\text{ d})\left(\frac{24\text{ h}}{1\text{ d}}\right)\left(\frac{3600\text{ s}}{1\text{ h}}\right)(4.90\times10^{-24})=46\text{ nm}$$

It would be 46 nm behind the photon.
Assess: This cosmic ray was traveling incredibly close to the speed of light. If you remember that violet light has a wavelength of around 400 nm and that to see something you have to use light with a wavelength smaller than the thing you are viewing, then we have that after traveling for a whole year, the photon and proton were so close together that if we could draw a very fine line at the location of the photon and another very fine line at the location of the proton, we would not be able to see the gap between the lines. It is not surprising, though, that the proton's speed was very close to c, considering how enormous its energy was: 48 J.

P27.67. Prepare: We will combine Equations 27.18 and 27.19 to find a relationship between energy, momentum, and velocity.

Solve: From Equation 27.18 $p=\gamma mu\Rightarrow\frac{p}{c}=\gamma m\frac{u}{c}\Rightarrow\frac{p}{c}=\gamma m\beta$.

Multiplying both sides by c^2, we have $pc=\gamma mc^2\beta$. Now using Equation 27.19, that is, $E=\gamma mc^2$, we simplify the momentum relationship to $pc=E\beta$. Thus,

$$p=\frac{\beta E}{c}=\frac{(0.95)(2.0\times10^{-10}\text{ J})}{3.0\times10^8\text{ m/s}}=6.3\times10^{-19}\text{ kg}\cdot\text{m/s}$$

P27.69. Prepare: We will use the relativistic expression for the total energy, Equation 27.19.
Solve: (a) The energy of the proton is

$$E = \gamma mc^2 = \frac{1}{\sqrt{1-v^2/c^2}}\,mc^2 = \frac{1}{\sqrt{1-(0.99)^2}}\,mc^2 = (7.089)(1.67\times10^{-27}\ \text{kg})(3.0\times10^8\ \text{m/s})^2$$

$$= 1.065\times10^{-9}\ \text{J}\times\frac{1\,\text{eV}}{1.6\times10^{-19}\ \text{J}} = 6.66\times10^9\ \text{eV} = 6660\ \text{MeV}$$

(b) Likewise, the energy of the electron is

$$E = \frac{1}{\sqrt{1-(0.99)^2}}(9.11\times10^{-31}\ \text{kg})(3.0\times10^8\ \text{m/s})^2 = 5.812\times10^{-13}\ \text{J}\times\frac{1\,\text{eV}}{1.6\times10^{-19}\ \text{J}} = 3.63\times10^6\ \text{eV} = 3.63\ \text{MeV}$$

P27.73. Prepare: The energy which can be obtained from a mass m is given by $E = mc^2$.
Solve: The chemical energy content of 1.0 kg of gasoline is $(46\ \text{MJ/kg})(1.0\ \text{kg}) = 46$ MJ. The mass which would be needed to generate this amount of energy, if the mass could be completely turned into energy is as follows:

$$m = \frac{E}{c^2} = \frac{46\times10^6\ \text{J}}{(3.0\times10^8\ \text{m/s})^2} = 5.1\times10^{-10}\ \text{kg}$$

Assess: A drop of gasoline this size would have a radius of about 0.06 mm. One way to interpret the result of this problem is to say that the mass of the gasoline decreases by 5.1×10^{-10} kg when it is combusted. Thus, in a sense, only 1 part in 2 billion ($\approx 5.1\times10^{-10}$) of the gasoline is fully used.

P27.77 Prepare: Mass and energy are equivalent and given by Equation 27.21.
Solve: (a) The mass "lost" in each reaction is

$$4m_\text{p} - m_\text{He} = 4(1.67\times10^{-27}\ \text{kg}) - 6.64\times10^{-27}\ \text{kg} = 0.04\times10^{-27}\ \text{kg}$$

In each fusion reaction, the energy released is

$$E_0 = mc^2 = (0.04\times10^{-27}\ \text{kg})\ (3.0\times10^8\ \text{m/s})^2 = 3.6\times10^{-12}\ \text{J} \approx 4\times10^{-12}\ \text{J}$$

(b) The initial mass of the four protons is 6.68×10^{-27} kg. The percent of mass lost during each fusion reaction is

$$\frac{0.04\times10^{-27}\ \text{kg}}{6.68\times10^{-27}\ \text{kg}}\times100 = 0.6\%$$

Since $m = E_0/c^2$, the energy released in one fusion reaction is also 0.6% of the initial rest energy.

28

QUANTUM PHYSICS

Q28.3. Reason: (a) For all materials that emit photo-electrons, the electrons need a minimum amount of energy in order to be released (called the work function). Since a single photon gives all its energy to a single electron, photons must have a minimum energy and hence frequency (called the threshold frequency).

(b) Increasing the light intensity increases the number of incident photons. Increasing the number of incident photons increases the number of electrons emitted. Since the moving electrons amount to an electric current, the more electrons, the more current.

(c) If the anode is positive by greater than about one volt, it attracts all electrons to the anode. A further increase in ΔV does not cause any more electrons to reach the anode and thus does not cause a further increase in the current I.

(d) As ΔV becomes more negative, fewer electrons have the energy to reach the anode. Hence a decrease in current.

(e) The greater the light intensity, the greater the number of photons reaching the photoelectron emitting surface. Since a single photon interacts with a single electron, the energy of the emitted electron depends on the energy of the single incident photon it interacts with, not the number of photons—it can only interact with one photon.

Classical physics can explain some aspects of the photoelectric effect, but, as mentioned in the text, not all of the aspects we see. Classically, we do not expect a sharp threshold frequency f_0. We also do not expect the current to appear instantly when the light is turned on. Classically, we expect that as the light intensity is increased, so is the energy absorbed by the surface electrons. Hence, the number of photoelectrons emitted, or the photocurrent, is expected to increase proportionately with light intensity. Classically, we expect the more intense light will give more energy to the photoelectrons and hence their stopping potential will be increases—this does not happen. Classical physics cannot explain A and E.

Assess: Only the quantum hypothesis can account for all the characteristics of the photoelectric effect. Einstein's explanation of the photoelectric effect won him a well-deserved Nobel Prize.

Q28.5. Reason: (a) A positive anode attracts all of the electrons to the anode. Once all of the electrons reach the anode, a further increase in ΔV does not cause any further increase in the current.

(b) If an electron has enough kinetic energy when it leaves the cathode it might just reach the anode. As ΔV decreases, fewer electrons are able to reach the anode until $\Delta V = -V_{\text{stop}}$ when no electrons reach the anode and the current stops.

(c) Electrons are still emitted, but they are not attracted to the anode and are in fact repelled by it. The electrons that are emitted do not have sufficient kinetic energy to reach the anode.

Assess: These explanations all fit with Einstein's quantum hypothesis.

Q28.9. Reason: (a) If the current is zero when $\Delta V = 1.0$ V that means the frequency of the light is less than the threshold frequency; $f < f_0$. So no electrons are ejected in the first place. Doubling the intensity of the light means doubling the number of photons hitting the metal, but the photons still aren't energetic enough to eject electrons; that is, the energy of the photons is less than the work function.

So no matter how many photons of this energy one shines on gold there won't be a photocurrent.

(b) Since no electrons are emitted in the first place, then increasing ΔV won't make any difference either.
Assess: Instead of computing energy it is also possible to think in terms of frequency and compare f to f_0.

Q28.11. Reason: The light quanta are discrete packets of energy. The fundamental unit of light energy is the photon. The energy is quantized.
Assess: This is equivalent to saying that energy is not continuously dividable. Many aspects of the universe appear to be discrete (chunky) rather than continuous.

Q28.13. Reason: When the wavelength of light is reduced by a factor of two, its frequency doubles, doubling the photon energy. If the intensity (proportional to the energy striking the cathode per second) is held constant, and each photon has twice as much energy, then half as many must strike the cathode per second as before. This means that half as many electrons will be emitted as before, so that the current will fall by a factor of two.
Assess: It makes sense that fewer but higher-energy photons lead to the same intensity of light, but that fewer photon mean fewer ejected electrons.

Q28.15. Reason: Both particles accelerate through the same potential difference and so have the same kinetic energy. Combine Equation 28.8 with $v = \sqrt{2K/m}$.

$$\lambda = \frac{h}{mv} = \frac{h}{m\sqrt{2K/m}} = \frac{h}{\sqrt{2Km}}$$

This shows that the particle with the smaller mass corresponds to the larger wavelength. Thus, the electron has the larger de Broglie wavelength.
Assess: Equation 28.8 shows that the wavelength is inversely proportional to the momentum.

Q28.19. Reason: The allowed energies of a particle in a box are given in Equation 28.13: $E_n = (hn)^2/(8mL^2)$.
(a) From the equation we see that for the energies to be large L, the length of the box should be very small.
(b) Based on the answer to part **(a)**, the system with the smallest dimension would be likely to have the largest energies; that is, the proton in the nucleus.
Assess: While it is true that the mass of the proton is larger than the mass of the electron and m is also in the denominator (which would argue the other way), the size difference is much greater than the mass difference, and the L is squared.

Q28.23. Reason: **(a)** It represents a changing speed because the wavelength is changing. The longer wavelength corresponds to a slower speed.
(b) The shorter wavelength on the left means the particle is moving faster on the left; the longer wavelength on the right means the particle is moving slower there.
Assess: This question reinforces the idea that faster particles have a shorter wavelength.

Q28.25. Reason: The uncertainty principle states that $\Delta x \Delta p \geq h/(2\pi)$. The uncertainty in the momentum can't be any larger than the momentum. When we put people back together, the momentum is zero or nearly zero—this means the uncertainty in the position is very large—people will be scrambled.
Assess: Movie makers (especially science fiction movie makers) need good technical advice in order to produce a correct product.

Q28.27. Reason: When f increases so does $E = hf$. And so will $K_{max} = E - E_0$.
The correct choice is B.
Assess: To get more electrons (choice A) we would need to increase the intensity of the light.

Q28.31. Reason: We must recall that $E = hf$, so the 100 MHz FM photons each have more energy than the 1000 kHz AM photons. Since both stations emit the same amount of energy each second, the AM station must emit more photons per second since each AM photon has less energy than one of the FM photons. The correct choice is B.
Assess: The two radio stations are not charged by the photon on their electric or power bill, but by how much energy they use, so their electric bills would be the same.

Q28.35. Reason: The relationship between energy and wavelength for a photo is $E = hc/\lambda$. From the diagram we see that $E_P > E_Q > E_R$, therefore $\lambda_P < \lambda_Q < \lambda_R$. The correct choice is A.

Assess: A photons wavelength is inversely proportional to its energy.

Problems

P28.1. Prepare: The angles of incidence for which diffraction from parallel planes occurs satisfy the Bragg condition. The Bragg condition is $2d\cos\theta_m = m\lambda$, where $m = 1, 2, 3, \ldots$

Solve: For first and second order diffraction,
$$2d\cos\theta_1 = (1)\lambda \qquad 2d\cos\theta_2 = (2)\lambda$$

Dividing these two equations,
$$\frac{\cos\theta_2}{\cos\theta_1} = 2 \Rightarrow \theta_2 = \cos^{-1}(2\cos\theta_1) = \cos^{-1}(2\cos 68°) = 41°$$

Assess: As m increases, the diffraction angle decreases as the equations show.

P28.3. Prepare: The angles corresponding to the various orders of diffraction satisfy the Bragg condition. The Bragg condition is $2d\cos\theta_m = m\lambda$, where $m = 1, 2, 3, \ldots$

Solve: The Bragg condition for $m = 1$ and $m = 2$ gives
$$2d\cos\theta_1 = (1)\lambda \qquad 2d\cos\theta_2 = (2)\lambda$$

Dividing these two equations,
$$\cos\theta_1 = \frac{\cos\theta_2}{2} = \frac{\cos 45°}{2} \Rightarrow \theta_1 = \cos^{-1}\left(\frac{\cos 45.0°}{2}\right) = 69.3°$$

Assess: For smaller values of m, the diffraction angle must increase.

P28.7. Prepare: Light of frequency f consists of discrete quanta, each of energy $E = hf = hc/\lambda$. The minimum energy required to free an electron is called the work function of the metal. An electron that has just absorbed a quantum of light energy can escape from the metal if its energy exceeds the work function.

Solve: The lowest photon energy that creates photoelectrons from the metal is
$$E = \frac{hc}{\lambda} = \frac{(6.63\times10^{-34}\,\text{J}\cdot\text{s})(3.0\times10^{8}\,\text{m/s})}{388\times10^{-9}\,\text{m}} \times \frac{1\,\text{eV}}{1.6\times10^{-19}\,\text{J}} = 3.20\,\text{eV}$$

This is the threshold energy. From Equation 28.4, the work function of the metal is $E_0 = 3.20\,\text{eV}$.

Assess: Electrons are able to absorb light quanta in the photoelectric effect.

P28.11. Prepare: The threshold frequency for the ejection of photoelectrons is $f_0 = E_0/h$, where E_0 is the work function.

Solve: The visible region of light extends from 400 nm to 700 nm. For $\lambda_0 = 400$ nm, the work function is
$$E_0 = f_0 h = \frac{hc}{\lambda_0} = \frac{(6.63\times10^{-34}\,\text{J}\cdot\text{s})(3.0\times10^{8}\,\text{m/s})}{400\times10^{-9}\,\text{m}} \times \frac{1\,\text{eV}}{1.6\times10^{-19}\,\text{J}} = 3.1\,\text{eV}$$

For $\lambda_0 = 700$ nm,
$$E_0 = \frac{(6.63\times10^{-34}\,\text{J}\cdot\text{s})(3.0\times10^{8}\,\text{m/s})}{700\times10^{-9}\,\text{m}} \times \frac{1\,\text{eV}}{1.6\times10^{-19}\,\text{J}} = 1.8\,\text{eV}$$

The cathode that will work in the entire visible range must have a work function of 1.8 eV or less.

Assess: A work function of 1.8 eV is reasonable. The greater the wavelength of the photon, the less the energy of the photon and hence it can only free electrons in a metal with a smaller work function.

P28.13. Prepare: We know the energy of a photon of frequency f and wavelength λ is $E = hf = hc/\lambda$.

Solve: (a) The longest wavelength is the photon with the minimum energy; longer wavelengths would not have enough energy to eject a photon. Solve the equation for λ.

$$\lambda = \frac{hc}{E} = \frac{(6.63\times10^{-34}\,\text{J}\cdot\text{s})(3.00\times10^{8}\,\text{m/s})}{4.3\,\text{eV}}\left(\frac{1\,\text{eV}}{1.60\times10^{-19}\,\text{J}}\right) = 2.9\times10^{-7}\,\text{m} = 290\,\text{nm}$$

(b) Review Example 28.4. If 4.3 eV of the energy is used just to get the electron ejected, the remaining 0.4 eV will become the kinetic energy of the electron, that is, $K_{max} = 0.4\,\text{eV}$.

$$v_{max} = \sqrt{\frac{2K_{max}}{m}} = \sqrt{\frac{2(0.4\,\text{eV})}{9.11\times10^{-31}\,\text{kg}}\left(\frac{1.60\times10^{-19}\,\text{J}}{1\,\text{eV}}\right)} = 3.7\times10^{5}\,\text{m/s}$$

Assess: Our data is a little different from that in Example 28.4, so we don't expect to get exactly the same answer; however we do get an answer that is in the same ballpark, so ours is probably correct.

P28.15. Prepare: See Example 28.4. The energy of the photons is $E = hf = hc/\lambda = (6.63\times10^{-34}\,\text{J}\cdot\text{s})(3.00\times10^{8}\,\text{m/s})/(350\,\text{nm}) = 5.68\times10^{-19}\,\text{J} = 3.55\,\text{eV}$. Because $V_{stop} = K_{max}/e$ we know that the maximum energy for the photoelectrons is $K_{max} = 1.25\,\text{eV}$.

Solve: (a) Assume the electrons are non-relativistic so that $K = 1/2mv^2$. Solve for v.

$$v = \sqrt{\frac{2K_{max}}{m}} = \sqrt{\frac{2(1.25\,\text{eV})}{9.11\times10^{-31}\,\text{kg}}\left(\frac{1.60\times10^{-19}\,\text{J}}{1\,\text{eV}}\right)} = 6.63\times10^{5}\,\text{m/s}$$

(b) Use Equation 28.6 to find the work function.

$$E_0 = hf - K_{max} = 3.55\,\text{eV} - 1.25\,\text{eV} = 2.30\,\text{eV}$$

Looking for this value in Table 28.1 tells us the metal must be potassium.

Assess: Our results seem to correspond to values in the examples, and we did arrive spot on for the work function of one of the metals listed in Table 28.1, so we probably did the problem correctly. The units all work out nicely.

P28.19. Prepare: The energy and wavelength of a photon are related by $E = hc/\lambda$. Before starting to solve the problem we might want to note that the arithmetic will be easier if we first calculate the quantity $hc(hc = 12.42\times10^{-7}\,\text{eV}\cdot\text{m})$.

Solve: Photon energies corresponding to the wavelengths given are

$$E = hc/\lambda = 12.42\times10^{-7}\,\text{eV}\cdot\text{m}/(4.37\times10^{-7}\,\text{m}) = 2.84\,\text{eV}$$

In like manner for $\lambda = 533\,\text{nm}$ obtain E = 2.33 eV, and for $\lambda = 564\,\text{nm}$ obtain E = 2.20 eV.

Assess: These are reasonable photon energies. Note that as the wavelength increases the energy decreases. This is consistent with the fact that the photon energy is inversely proportional to the wavelength.

P28.23. Prepare: Follow Example 28.5 and use Equation 28.3. We are given $f = 870\,\text{kHz}$ and $P = 50{,}000\,\text{W}$.

Solve:

$$R = \frac{P}{hf} = \frac{50{,}000\,\text{W}}{(6.63\times10^{-34}\,\text{J}\cdot\text{s})(870\times10^{3}\,\text{Hz})} = 8.7\times10^{31}\,\text{photons per second}$$

Assess: This result is much larger than the answer in Example 28.5 because the power of the radio station is much larger than the power output of a laser pointer and because the frequency of the radio photons is so much less than the frequency of the visible red laser photons.

P28.27. Prepare: Current is related to the number of electrons per unit time by

$$I = \frac{Q}{t} = \frac{Ne}{t} = e\left(\frac{N}{t}\right)_{\text{electrons}}$$

The number of electrons per unit time is the same as the number of photons per unit time

$$\left(\frac{N}{t}\right)_{\text{electrons}} = \left(\frac{N}{t}\right)_{\text{photons}}$$

Power of the light flash is related to the number of photons per unit time by

$$P = \frac{E_{\text{total}}}{t} = \frac{NE_{\text{photon}}}{t} = \left(\frac{N}{t}\right)_{\text{photons}} \frac{hc}{\lambda}$$

Solve: Combine these to obtain the power of the light flash

$$P = \left(\frac{N}{t}\right)_{\text{photons}} \frac{hc}{\lambda} = \left(\frac{N}{t}\right)_{\text{electrons}} \frac{hc}{\lambda} = \frac{I}{e} \frac{hc}{\lambda} = \frac{(4.2 \times 10^{-7}\,\text{A})(19.89 \times 10^{-26}\,\text{J} \cdot \text{m})}{(1.60 \times 10^{-19}\,\text{C})(5.50 \times 10^{-7}\,\text{m})} = 9.5 \times 10^{-7}\,\text{W}$$

Assess: Since the detected current is in the micro-Ampere range, we expect a small power rating for the flash.

P28.29. Prepare: The de Broglie wavelength for a moving particle is given by Equation 28.8, $\lambda = h/p = h/(mv)$. The baseball's momentum is $p = mv = (0.200\,\text{kg})(30\,\text{m/s}) = 6.0\,\text{kg m/s}$.

Solve: **(a)** The baseball's de Broglie wavelength is

$$\lambda = \frac{h}{p} = \frac{6.63 \times 10^{-34}\,\text{J} \cdot \text{s}}{6.0\,\text{kg} \cdot \text{m/s}} = 1.1 \times 10^{-34}\,\text{m}$$

(b) Using $\lambda = h/p = h/(mv)$, we have

$$v = \frac{h}{m\lambda} = \frac{6.63 \times 10^{-34}\,\text{J} \cdot \text{s}}{(0.200\,\text{kg})(0.20 \times 10^{-9}\,\text{m})} = 1.7 \times 10^{-23}\,\text{m/s}$$

Assess: This velocity is too small to be measured.

P28.31. Prepare: The kinetic energy of a particle of mass m is related to its momentum by $E = p^2/(2m)$. The momentum of the particle is related to its wavelength through $\lambda = h/p$.

Solve: Using $\lambda = h/p$, for an electron we have

$$E = \frac{p^2}{2m} = \frac{h^2}{2m\lambda^2} = \frac{(6.63 \times 10^{-34}\,\text{J} \cdot \text{s})^2}{2(9.11 \times 10^{-31}\,\text{kg})(1.0 \times 10^{-9}\,\text{m})^2} = 2.4 \times 10^{-19}\,\text{J}$$

In eV, the energy is $(2.41 \times 10^{-19}\,\text{J})\left(\dfrac{1\,\text{eV}}{1.6 \times 10^{-19}\,\text{J}}\right) = 1.5\,\text{eV}$.

Assess: This is a small but reasonable amount of energy for an electron.

P28.33. Prepare: From the given de Broglie wavelength of 10 fm, we will first determine the speed of the proton and then its kinetic energy. Also note the conversion: $1\,\text{eV} = 1.6 \times 10^{-19}\,\text{J}$.

Solve: The de Broglie wavelength is $\lambda = h/(mv)$. Thus

$$v = \frac{h}{m\lambda} = \frac{6.63 \times 10^{-34}\,\text{J} \cdot \text{s}}{(1.67 \times 10^{-27}\,\text{kg})(10 \times 10^{-15}\,\text{m})} = 3.970 \times 10^{7}\,\text{m/s}$$

So, the kinetic energy of the proton is

$$K = \frac{1}{2}mv^2 = \frac{1}{2}(1.67 \times 10^{-27}\,\text{kg})(3.970 \times 10^{7}\,\text{m/s})^2 \times \frac{1\,\text{eV}}{1.6 \times 10^{-19}\,\text{J}} = 8.2\,\text{MeV}$$

Assess: Since the proton has such a high speed, we expect it to have a large kinetic energy.

P28.39. Prepare: We will model the 5.0-fm-diameter nucleus as a box of length $L = 5.0$ fm. Equation 28.13 tells us that the energy of the confined proton is

$$E_n = \frac{h^2}{8mL^2}n^2 \qquad n = 1, 2, 3, 4,\ldots$$

Solve: The proton's energy is restricted to the discrete values

$$E_n = \frac{h^2}{8mL^2}n^2 = \frac{(6.63 \times 10^{-34}\,\text{J} \cdot \text{s})^2 n^2}{8(1.67 \times 10^{-27}\,\text{kg})(5.0 \times 10^{-15}\,\text{m})^2} = (1.316 \times 10^{-12}\,\text{J})n^2$$

where $n = 1, 2, 3, \ldots$ For $n = 1$, $E_1 = 1.316 \times 10^{-12}\,\text{J} = 1.3 \times 10^{-12}\,\text{J}$, for $n = 2$, $E_2 = (1.316 \times 10^{-12}\,\text{J})4 = 5.3 \times 10^{-12}\,\text{J}$, and for $n = 3$, $E_3 = 9E_1 = 1.2 \times 10^{-11}\,\text{J}$.

Assess: Energies increase with n^2 as we calculated.

P28.41. Prepare: To conserve energy, the emission and the absorption photons must have exactly the energy lost or gained by the quantum system in the appropriate quantum jumps. We will use Equation 28.16, which tells us that $E_2 - E_1 = hf = hc/\lambda$.

Solve: The wavelength of the emission photon from the $n = 2$ to $n = 1$ transition is

$$\lambda = \frac{hc}{E_2 - E_1} = \frac{(4.14 \times 10^{-15} \text{ eV s})(3.0 \times 10^8 \text{ m/s})}{1.5 \text{ eV}} = 830 \text{ nm}$$

Likewise, $\lambda = 500$ nm for the $3 \rightarrow 2$ transition with $\Delta E = 2.5$ eV, and $\lambda = 310$ nm for the $3 \rightarrow 1$ transition with $\Delta E = 4.0$ eV.

Assess: These are all reasonable wavelengths.

P28.45. Prepare: The position uncertainty and the momentum uncertainty are related by $\Delta x \Delta p_x \geq \hbar$. The momentum, velocity, and mass of an object are related by $p_x = mv_x$.

Solve: The uncertainty in the position of the object is

$$\Delta x \geq \hbar/\Delta p_x = \hbar/(m\Delta v_x) = (1.05 \times 10^{-34} \text{ J} \cdot \text{s})/((9.11 \times 10^{-31} \text{ kg})(0.2 \times 10^6 \text{ m/s})) = 5.8 \times 10^{-10} \text{ m}$$

Assess: This seems like a small uncertainty in position to us, but to an electron it is very large. In fact it is 10 times larger than the Bohr radius.

P28.47. Prepare: The Heisenberg uncertainty principle is given in Equation 28.17.

$$\Delta x \Delta p_x = \Delta x \, m \Delta v_x \geq \frac{h}{2\pi}$$

We are given $\Delta x = 1 \times 10^{-4}$ m and $\rho = 1000$ kg/m^3. A preliminary calculation will give us m. We are also given $r = d/2 = 25 \times 10^{-9}$ m.

$$m = \rho V = \rho \frac{4}{3}\pi r^3 = (1000 \text{ kg/m}^3)\left(\frac{4}{3}\right)\pi(25 \times 10^{-9} \text{ m})^3 = 6.5 \times 10^{-20} \text{ kg}$$

Solve: Solve for the uncertainty in speed.

$$\Delta v_x \geq \frac{h}{2\pi m \Delta x} = \frac{6.63 \times 10^{-34} \text{ J} \cdot \text{s}}{2\pi(6.5 \times 10^{-20} \text{ kg})(1 \times 10^{-4} \text{ m})} = 1.6 \times 10^{-11} \text{ m/s}$$

Assess: This uncertainty in speed is quite small because its uncertainty in position is relatively large.

P28.51. Prepare: The threshold energies for potassium and gold are given in Table 28.1.

Solve: (a) The threshold frequency is $f_0 = E_0/h$. The threshold frequency for potassium and gold are given in the table in part **(b)**.

(b) The corresponding threshold wavelength is $\lambda_0 = c/f_0$. The results of the calculations are in the following table.

Metal	E_0 (eV)	f_0 (Hz)	λ_0 (nm)
Potassium	2.30	5.56×10^{14}	540
Gold	5.10	1.23×10^{15}	244

(c) When light of wavelength λ is incident on the metal, the maximum kinetic energy of the photoelectrons is given by Equation 28.6:

$$K_{max} = \frac{1}{2}mv_{max}^2 = hf - E_0 = \frac{hc}{\lambda} - E_0 \Rightarrow v_{max} = \sqrt{\frac{2}{m}\left(\frac{hc}{\lambda} - E_0\right)}$$

E_0 must be converted to SI units of joules before this formula can be used. v_{max} for potassium and gold are given in the table in part **(d)**.

(d) The stopping potential from Equation 28.7 is

$$V_{stop} = \frac{K_{max}}{e} = \frac{1}{e}\left(\frac{hc}{\lambda} - E_0\right)$$

where again E_0 has to be joules. The results of the calculations are in the following table.

Metal	E_0 (J)	v_{max} (m/s)	V_{stop} (V)
Potassium	3.68×10^{-19}	10.8×10^5	3.35
Gold	8.16×10^{-19}	4.40×10^5	0.55

Assess: Values obtained for the threshold frequency, threshold wavelength, maximum speed of the emitted photoelectrons, and stopping potential are reasonable.

P28.53. Prepare: We will use Equation 28.7.
Solve: The stopping potential is $eV_{stop} = hf - E_0$. Substituting the given values,

$$eV_{stop} = \frac{hc}{300\text{ nm}} - E_0 \quad \text{and} \quad 0.257(eV_{stop}) = \frac{hc}{400\text{ nm}} - E_0$$

Substituting the first equation into the second,

$$0.257\left(\frac{hc}{300\text{ nm}} - E_0\right) = \frac{hc}{400\text{ nm}} - E_0 \Rightarrow E_0(1 - 0.257) = hc\left(\frac{1}{400\text{ nm}} - \frac{0.257}{300\text{ nm}}\right)$$

or

$$E_0 = \frac{(4.14\times10^{-15}\text{ eV}\cdot\text{s})(3.0\times10^8\text{ m/s})(1.643\times10^{-3}/\text{nm})}{0.743} = 2.75\text{ eV}$$

From Table 28.1, we identify the cathode metal as sodium.
Assess: A work function of 2.75 eV is reasonable for metal and Table 28.1 allows us to identify the metal as sodium.

P28.55. Prepare: Equation 28.8 gives the de Broglie wavelength for a moving object.
We are given $m = 1.00\times10^{-14}$ kg and $v = 0.400\times10^{-2}$ m/s.
Solve:

$$\lambda = \frac{h}{mv} = \frac{6.63\times10^{-34}\text{ J}\cdot\text{s}}{(1.00\times10^{-14}\text{ kg})(0.400\times10^{-2}\text{ m/s})} = 1.66\times10^{-17}\text{ m}$$

This wavelength is so tiny (much smaller than the atoms the blood cell is made of) that we do not need to consider the wave nature of blood cells when describing the flow of blood in the body.
Assess: The wavelength is so tiny because h is so small. The world is quantized, but the chunks are very small. If h were much bigger then quantum physics would be common sense to us.
We report the answer to three significant figures as indicated by the data. Appropriate units cancel to leave the answer in meters.

P28.57. Prepare: Photons of frequency f are particles with energy given as $E = hf = hc/\lambda$.
Solve: (a) The wavelength is calculated as follows:

$$E_{gamma} = hf = h\left(\frac{c}{\lambda}\right) \Rightarrow \lambda = \frac{(6.63\times10^{-34}\text{ J}\cdot\text{s})(3.0\times10^8\text{ m/s})}{1.0\times10^{-13}\text{ J}} = 2.0\times10^{-12}\text{ m}$$

(b) The energy of a visible-light photon of wavelength 500 nm is

$$E_{visible} = h\left(\frac{c}{\lambda}\right) = \frac{(6.63\times10^{-34}\text{ J}\cdot\text{s})(3.0\times10^8\text{ m/s})}{500\times10^{-9}\text{ m}} = 3.978\times10^{-19}\text{ J}$$

The number of photons n such that $E_{gamma} = nE_{visible}$ is

$$n = \frac{E_{gamma}}{E_{visible}} = \frac{1.0\times10^{-13}\text{ J}}{3.978\times10^{-19}\text{ J}} = 2.51\times10^5$$

Assess: Gamma ray photons have higher energy compared to visible photons.

P28.63. Prepare: The energy of a photon is hc/λ. We are given $\lambda = 532$ nm, $\Delta t = 100$ ps, and $A = \pi r^2 = \pi(0.020\text{ m})^2 = 1.256\text{ m}^2$.

Solve:

$$I = \frac{P}{A} = \frac{E/\Delta t}{A} = \frac{nE_{\text{photon}}/\Delta t}{A} = \frac{n\frac{hc}{\lambda}/\Delta t}{A} \Rightarrow$$

$$n = \frac{IA\Delta t\lambda}{hc} = \frac{(300\text{ W/m}^2)(0.001257\text{ m}^2)(100\text{ ps})(532\text{ nm})}{(6.63\times10^{-34}\text{ J}\cdot\text{s})(3.00\times10^8\text{ m/s})} = 1.0\times10^8\text{ photons}$$

Assess: In some contexts this is a big number, but it isn't a lot of photons. Many fewer than this make it back to the detector on earth.

P28.65. Prepare: The energy of a photon is related to its wavelength by $E = hc/\lambda$. The total energy is related to the number of photons and the energy per photon by $E = NE_{\text{photon}}$.

Solve: (a) The energy of the photon with wavelength 2.54×10^{-7} m is

$$E = hc/\lambda = 19.89\times10^{-26}\text{ J}\cdot\text{m}/(2.54\times10^{-7}\text{ m}) = 7.83\times10^{-19}\text{ J}$$

In like manner, the energy of the photon with wavelength 3.00×10^{-7} m is 6.63×10^{-19} J.

(b) The number of photons for the first case is

$$N = E/E_{\text{photon}} = 3.7\times10^{-3}\text{ J}/(7.83\times10^{-19}\text{ J}) = 4.7\times10^{15}\text{ photons}$$

In like manner for the second case obtain $N = 2.0\times10^{16}$ photons.

(c) A sunburn will require more of the less energetic photons.

Assess: We expect a large number of photons because the energy of each photon is very small.

P28.69. Prepare: Recall the equation for the diffraction through a circular aperture from Chapter 17 (we make the small angle approximation here).

$$w = \frac{2.44\lambda L}{D} \Rightarrow \lambda = \frac{wD}{2.44L}$$

where λ is the de Broglie wavelength of the electrons and $\lambda = h/mv$.

Solve: Solve for the speed of the electrons

$$v = \frac{h}{m\frac{wD}{2.44L}} = \frac{2.44hL}{mwD}$$

Now the kinetic energy is

$$E = \frac{1}{2}mv^2 = \frac{1}{2}m\left(\frac{2.44hL}{mwD}\right)^2$$

$$= \frac{1}{2}(9.11\times10^{-31}\text{ kg})\left(\frac{2.44(6.63\times10^{-34}\text{ J}\cdot\text{s})(1.2\text{ m})}{(9.11\times10^{-31}\text{ kg})(0.37\text{ mm})(200\text{ nm})}\right)^2\left(\frac{1\text{ eV}}{1.6\times10^{-19}\text{ J}}\right) = 2.4\text{ keV}$$

Assess: This is a typical energy for electrons.

P28.71. Prepare: The de Broglie wavelength is given in Equation 28.8: $\lambda = h/p$.

An electron accelerated through a potential difference of $1.00\times10^5 V$ gains kinetic energy of

$$K = 1.00\times10^5\text{ eV}\left(\frac{1.60\times10^{-19}\text{ J}}{1\text{ eV}}\right) = 1.60\times10^{-14}\text{ J}$$

Solve: (a) The classical expressions for $K = 1/2mv^2$ and $p = mv$ give

$$\lambda = \frac{h}{p} = \frac{h}{mv} = \frac{h}{m\sqrt{2K/m}} = \frac{h}{\sqrt{2Km}} = \frac{6.63\times10^{-34}\,\text{J}\cdot\text{s}}{\sqrt{2(1.60\times10^{-14}\,\text{J})(9.11\times10^{-31}\,\text{kg})}} = 3.88\times10^{-12}\,\text{m}$$

(b) We first solve for γ using the relativistic expression for K: $K = (\gamma-1)mc^2$.
Solve this for γ.

$$\gamma = \frac{K}{mc^2} + 1 = \frac{1.60\times10^{-14}\,\text{J}}{(9.11\times10^{-31}\,\text{kg})(3.00\times10^8\,\text{m/s})^2} + 1 = 1.1951$$

We then find u from the definition of γ: $\gamma = 1/\sqrt{1-\frac{u^2}{c^2}}$.

Solve this for u.

$$u = c\sqrt{1 - \frac{1}{\gamma^2}} = (3.00\times10^8\,\text{m/s})\sqrt{1 - \frac{1}{1.1951^2}} = 1.6429\times10^8\,\text{m/s}$$

Finally, the relativistic expression for $p = \gamma mu$ gives

$$\lambda = \frac{h}{p} = \frac{h}{\gamma mu} = \frac{6.63\times10^{-34}\,\text{J}\cdot\text{s}}{(1.1951)(9.11\times10^{-34}\,\text{J}\cdot\text{s})(1.6429\times10^8\,\text{m/s})} = 3.71\times10^{-12}\,\text{m.}$$

Assess: The electron is going fast enough that relativity does make a small difference in the de Broglie wavelength. The relativistic wavelength is shorter than the one calculated classically.

P28.73. Prepare: A particle confined in a one-dimensional box has discrete energy levels given by Equation 28.13.
Solve: (a) Equation 28.13 for the $n = 1$ state is

$$E_n = \frac{h^2}{8mL^2} = \frac{(6.63\times10^{-34}\,\text{J}\cdot\text{s})^2}{8(2.7\times10^{-3}\,\text{kg})(0.10\,\text{m})^2} = 2.04\times10^{-63}\,\text{J}$$

The minimum energy of the Ping-Pong ball is $E_1 = 2.0\times10^{-63}$ J.
(b) The speed is calculated as follows:

$$E_1 = 2.04\times10^{-63}\,\text{J} = \frac{1}{2}mv^2 = \frac{1}{2}(2.7\times10^{-3}\,\text{kg})v^2 \Rightarrow v = \sqrt{\frac{2(2.04\times10^{-63}\,\text{J})}{2.7\times10^{-3}\,\text{kg}}} = 1.2\times10^{-30}\,\text{m/s}$$

Assess: Such small energies and velocities are not measurable.

P28.77. Prepare: The relationship between the energy of an electron in a box and the size of the box is

$$E = h^2 n^2 / 8mL^2 \quad \text{where} \quad n = 1, 2, 3, 4, \ldots$$

Solve: Solving the above expression for the length of the box and noting that the smallest box occurs for the case of $n = 1$, obtain

$$L = h/\sqrt{8mE}$$

Inserting $E = 1.28\,\text{eV}$, obtain the length of smallest box for that electron

$$L = (6.63\times10^{-34}\,\text{J}\cdot\text{s})/\sqrt{8(9.11\times10^{-31}\,\text{kg})(1.28\,\text{eV})(1.6\times10^{-19}\,\text{J/eV})} = 5.43\times10^{-10}\,\text{m}$$

Standing wave resonance may be established for the electron in a box of this length or any multiple of this length.
Inserting $E = 1.88\,\text{eV}$, obtain the length of smallest box for that electron

$$L = (6.63\times10^{-34}\,\text{J}\cdot\text{s})/\sqrt{8(9.11\times10^{-31}\,\text{kg})(2.88\,\text{eV})(1.6\times10^{-19}\,\text{J/eV})} = 3.62\times10^{-10}\,\text{m}$$

Standing wave resonance may be established for the electron in a box of this length or any multiple of this length.
The length of the box must be such that it can accommodate standing wave resonance for either electron. This means that it is some multiple (say p) of 5.43×10^{-10} m and some other multiple (say q) of 3.62×10^{-10} m.

This allows us to write $(5.43\times10^{-10}\,\text{m})\text{p} = (3.62\times10^{-10}\,\text{m})\text{q}$

Solving for q/p obtain $\text{q/p} = 3/2$.

Then note that $(5.43\times10^{-10}\,\text{m})\text{p} = (5.43\times10^{-10}\,\text{m})(2) = 10.9\times10^{-10}\,\text{m}$

Also note that $(3.62\times10^{-10}\text{ m})q = (3.62\times10^{-10}\text{ m})(3) = 10.9\times10^{-10}$ m

The shortest box that both electrons can establish standing wave resonance in is $L = 10.9\times10^{-10}$ m.

Assess: A critical piece of information needed for the solution of this problem is that both electrons must be able to establish standing wave resonance in the same box. This means the box must be a multiple of the length established for the n = 1 case.

P28.79. Prepare: The allowed energies of a particle of mass m in a two-dimensional square box of side L are

$$E_{nm} = \frac{h^2}{8mL^2}(n^2 + l^2)$$

The two quantum numbers, n and l, take integer values 1, 2, 3, …
Solve: (a) The minimum energy for a particle is for $n = l = 1$:

$$E_{min} = E_{11} = \frac{h^2}{8mL^2}(1^2 + 1^2) = \frac{h^2}{4mL^2}$$

(b) The five lowest allowed energies are E_{min}, $(5/2)E_{min}$ (for $n = 1$, $l = 2$ and $n = 2$, $l = 1$), $4E_{min}$ (for $n = 2$, $l = 2$), $5E_{min}$ (for $n = 1$, $l = 3$ and $n = 3$, $l = 1$), and $(13/2)E_{min}$ (for $n = 2$, $l = 3$ and $n = 3$, $l = 2$).
Assess: Several different combinations give the same energy, e.g. n = 1, $l = 2$ and $n = 2$ and $l = 1$.

P28.83. Prepare: The energy of a photon is related to the frequency by $E = hf$. In Integrated Example 28.16 you will find that the amount of energy needed to "flip" a proton of magnetic moment $\mu = 1.41\times10^{-26}$ J/T in a magnetic of strength B is $\Delta E = 2\mu B$. Finally, the necessary frequency for the probe coil is the same as the frequency of a photon with sufficient energy to cause the electron to "flip."
Solve: Combine these and solve for the magnetic field.

$$hf = 2\mu B \Rightarrow B = \frac{hf}{2\mu} = \frac{(6.63\times10^{-34}\text{ J}\cdot\text{s})(8.00\times10^{8}\text{ Hz})}{2(1.41\times10^{-26}\text{ J/T})} = 19\text{ T}$$

Assess: This is a large but doable magnetic field.

ATOMS AND MOLECULES

Q29.1. Reason: The red-orange colors in the neon emission spectrum are due to transitions from excited $3p$ states to the lower energy but still excited $3s$ states. This occurs because the ground states are collisionally excited by the electrical discharge. The absorption spectrum of a gas consists of *only* those spectral lines that start from the ground state. This is because essentially all the atoms are in the ground state and the population of excited states is too small for absorption from these states to be noticed. For neon in a glass tube, there's no population of atoms in the $3s$ state to absorb red-orange light. The only wavelengths that can be absorbed are those absorbed by ground-state neon atoms. The lowest excited states are greater than 16 eV above the ground state, so all absorption lines in neon will have wavelengths less than 100 mm. Neon is transparent to all visible wavelengths of light.

Assess: This underscores why absorption wavelengths are a subset of the emission wavelengths.

Q29.3. Reason: Photons are absorbed in a quantum jump from a lower energy level to a higher energy level. Because absorption wavelengths are a subset of the wavelengths in the emission spectrum, we can use Equation 29.17 to analyze the hydrogen absorption spectrum. Because most of the atoms in a gas are in the ground state ($n=1$), the only quantum jumps seen in the absorption spectrum start from this state. In other words, only the Lyman series ($m=1$ and $n=m+1,\ m+2,\dots$) is observed in the absorption spectrum. Using $m=1$ and $n=2,3,4,\dots$) in Equation 29.17, we do not find any spectral line with $\lambda=656$ nm. However, a transition $2\to 3$ does lead to a line with $\lambda=656$ nm. However, this line is in the emission spectrum, not the absorption spectrum.

Assess: The absorption spectrum must generally have the lower of the two quantum levels be the ground state $n=1$.

Q29.9. Reason: The number indicates the value of n, and the letter indicates the value of l.

(a) $n=4$ $l=2$

(b) $n=5$ $l=3$

(c) $n=6$ $l=0$

Assess: Remember that the values of l begin with 0 while the values of n begin with 1.

Q29.11. Reason: Acceptable values for l are $l=0,1,2,3,\dots,n-1$. So if $n=5$, then the values for l are 0, 1, 2, 3, 4.

Assess: The values for l start with 0.

Q29.13. Reason: From the picture of the periodic table in Figure 29.23 we see that elements 87 and 88, which correspond to the $7s$ energy level, appear before the actinides (elements 89 through 102) which correspond to the $5f$ energy level. Therefore a $7s$ electron has lower energy than a $5f$ electron.

Assess: These interesting anomalies are why the periodic table has its characteristic saddle shape.

Q29.15. Reason: The ground state configuration for a multi-electron atom is given by

$$1s^2 2s^2 2p^6 3s^2 3p^6 4s^2 3d^{10} 4p^6 5s^2 4d^{10} 5p^6 6s^2 4f^{14} 5d^{10} 6p^6 \ldots$$

Note that that we have $3d^{10}$, $4d^{10}$, $5d^{10}$, etc. This informs us that it takes ten electrons to fill these subshells. For the d subshell, the orbital quantum number is $l = 2$, there are $2l + 1$ values of the magnetic quantum number m (2, 1, 0, -1 and -2) and each of these can have $m_s = \pm(1/2)$. Hence a total of ten electrons are needed to fill these subshells.

Assess: The ten elements that provide these ten electrons are called transition elements.

Q29.21. Reason: Since the wavelengths are labeled with subscripts $1 - 7$, let's agree to refer to these as transition 1 through 7. Transition 1 is associated with λ_1, etc.

Transition 1 is not allowed because it violates the selection rule $\Delta l = \pm 1$.
Transition 2 is in the emission spectrum and it is just out of the visible range at the blue end and into the UV.
Transition 3 is not allowed because it violates the selection rule $\Delta l = \pm 1$.
Transition 4 is not allowed because it violates the selection rule $\Delta l = \pm 1$.
Transition 5 is in the IR range of the emission spectrum.
Transition 6 is in the IR range of the emission spectrum.
Transition 7 is in the IR range of the emission spectrum.
Assess: An energy level diagram provides a tremendous amount of information.

Q29.23. Reason: In these atoms we have a population inversion ($N_3 > N_2$) and we clearly have a mechanism for getting more atoms into level 3 than into level 2. Next note that the transition from level 3 to level 2 is allowed ($\Delta l = +1$). If one atom undergoes spontaneous emission from level 3 to level 2, we will have a photon with the exact amount of energy needed to start the laser action.
Assess: All requirements needed for a laser appear to have been met.

Q29.25. Reason: If $l = 4$, there are $2l + 1$ or nine possible values for the magnetic quantum number ($m = 4, 3, 2, 1, 0, -1, -2, -3$, and -4). Each of these states can have $m_s = \pm(1/2)$ so this provides a total of 18 states.
The correct choice is D.
Assess: The quantum numbers provide a lot of information.

Q29.29. Reason: The angular momentum of an orbiting electron is determined by $L = n\hbar$. The energy levels of the stationary states of the hydrogen atom are determined by $E_n = -13.6\,\text{eV}/n^2$. We will solve the first expression for the principle quantum number n and then use n to determine the energy of the atom.
The principle quantum number is

$$n = L/\hbar = (3.18 \times 10^{-34}\,\text{kg} \cdot \text{m}^2/\text{s})/(1.06 \times 10^{-34}\,\text{J} \cdot \text{s}) = 3$$

The energy of a Bohr hydrogen atom in the $n = 3$ state is

$$E_n = -13.6\,\text{eV}/n^2 = -13.6\,\text{eV}/9 = -1.51\,\text{eV}$$

The correct response is D.
Assess: An acceptable energy level state is $n = 3$ and as shown in Table 29.2, when the atom is in this state the energy is -1.51 eV.

Problems

P29.3. Prepare: We'll look at the emission spectrum, so the transitions we are interested in are $4 \rightarrow 3, 5 \rightarrow 3$, and $6 \rightarrow 3$.
Equation 29.2 gives the generalized Balmer formula.

$$\lambda_{n \rightarrow 3} = \frac{91.1\,\text{nm}}{\left(\frac{1}{9} - \frac{1}{n^2}\right)}$$

Solve: Plug in 4, 5, and 6 for n.

$$\lambda_{4\to3} = \frac{91.1\,\text{nm}}{\left(\frac{1}{9} - \frac{1}{16}\right)} = 1870\,\text{nm}$$

$$\lambda_{5\to3} = \frac{91.1\,\text{nm}}{\left(\frac{1}{9} - \frac{1}{25}\right)} = 1280\,\text{nm}$$

$$\lambda_{6\to3} = \frac{91.1\,\text{nm}}{\left(\frac{1}{9} - \frac{1}{36}\right)} = 1090\,\text{nm}$$

These wavelengths are in the infrared (IR) region of the spectrum.

Assess: The Lyman series wavelengths are in the UV region of the spectrum, the Balmer series in the visible, and the Paschen series in the IR region.

P29.5. Prepare: For a neutral atom, the number of electrons is the same as the number of protons. The atomic number of Be, C, and N is 4, 6, and 7, respectively.

Solve: (a) Since $Z = 4$, the $^9\text{Be}^+$ ion has 3 electrons, 4 protons, and $9 - 4 = 5$ neutrons.

(b) Since $Z = 6$, ^{12}C has 6 electrons, 6 protons, and $12 - 6 = 6$ neutrons.

(c) Since $Z = 7$, the triply charged $^{15}\text{N}^{+++}$ ion has 4 electrons, 7 protons, and $15 - 7 = 8$ neutrons.

Assess: We need a good understanding of the atomic model to label atoms and ions.

P29.9. Prepare: To conserve energy, the absorption spectrum must have exactly the energy gained by the atom in the quantum jumps.

Solve: (a) An electron with a kinetic energy of 2.0 eV can collide with an atom in the $n = 1$ state and raise its energy to the $n = 2$ state. This is possible because $E_2 - E_1 = 1.5\,\text{eV}$ is less than 2.0 eV. On the other hand, the atom cannot be excited to the $n = 3$ state.

(b) The atom will absorb 1.5 eV of energy from the incoming electron, leaving the electron with 0.50 eV of kinetic energy.

P29.11. Prepare: We'll first compute the kinetic energy of an electron traveling at $1.6 \times 10^6\,\text{m/s}$.

Solve:

$$K = \frac{1}{2}mv^2 = \frac{1}{2}(9.11\times10^{-31}\,\text{kg})(1.6\times10^6\,\text{m/s})^2\left(\frac{1\,\text{eV}}{1.6\times10^{-19}\,\text{J}}\right) = 7.29\,\text{eV}$$

If the electron excites the 6.0 eV level than there is $7.29\,\text{eV} - 6.0\,\text{eV} = 1.29\,\text{eV}$ left. This equates to a speed of

$$v = \sqrt{\frac{2K}{m}} = \sqrt{\frac{2(1.29\,\text{eV})\left(\frac{1.6\times10^{-19}\,\text{J}}{1\,\text{eV}}\right)}{9.11\times10^{-31}\,\text{kg}}} = 6.7\times10^5\,\text{m/s}$$

This is the minimum speed the electron could have after the collision. A similar calculation shows that when the 4.0 eV level is excited the maximum speed is $1.1\times10^6\,\text{m/s}$.

Assess: These are typical electron speeds.

P29.13. Prepare: In the Bohr model of the hydrogen atom, an integer number of electron wavelengths fit around the circumference of the electron orbit. We also know the angular momentum is quantized: $L_n = p_n r_n = n\hbar \Rightarrow p_n = n\hbar/r_n$.

Solve: The de Broglie wavelength is $\lambda_n = h/p_n$. Substitute for p_n and recall that $\hbar = h/(2\pi)$.

$$\lambda_n = \frac{hr_n}{n\hbar} = \frac{2\pi r_n}{n}$$

For $n = 3$, $3\lambda_3 = 2\pi r_3$. $2\pi r_3$ is the circumference of the $n = 3$ orbit, so three electron wavelengths fit around the this orbit.

Assess: Of course, this would mean six nodes and six antinodes for $n = 3$. See Figure 29.16.

P29.15. Prepare: $E = \dfrac{hc}{\lambda} = \dfrac{(4.14\times10^{-15}\,\text{eV}\cdot\text{s})(3.0\times10^{8}\,\text{m/s})}{1870\,\text{nm}} = 0.66\,\text{eV}$

Solve: We see that this energy difference is the gap between states 4 and 3 as the electron drops down.
-0.85 eV $- (-1.51$ eV$) = 0.66$ eV, so the transition is the $4 \rightarrow 3$ transition.

Assess: These are small integers and quite reasonable.

P29.19. Prepare: We will use Equation 29.2 or 29.17 with $\lambda_0 = 91.1$ nm.

Solve: Photons emitted from the $n = 4$ state start in energy level $n = 4$ and undergo a quantum jump to a lower energy level with $m < 4$. The possibilities are $4 \rightarrow 1, 4 \rightarrow 2$, and $4 \rightarrow 3$. According to Equation 29.17, the transition $4 \rightarrow m$ emits a photon of wavelength

$$\lambda = \frac{\lambda_0}{(1/m^2 - 1/n^2)} = \frac{91.1\,\text{nm}}{(1/m^2 - 1/16)}$$

These values are given in the following table.

Transition	Wavelength
$4 \rightarrow 1$	97.2 nm
$4 \rightarrow 2$	486 nm
$4 \rightarrow 3$	1870 nm

Assess: Transitions are in the ultraviolet, visible, and infrared regions.

P29.23. Prepare: Because -13.60 eV $+ 12.09$ eV $= -1.51$ eV, we know the atom is in the $n = 3$ state from Table 29.2.

Solve: Equation 29.19 says $L = \sqrt{l(l+1)}\,\hbar$ $\quad l = 0, 1, 2, \ldots, n-1$, so the maximum angular momentum is when $l = 2$ and it is $L = \sqrt{2(2+1)}\,\hbar = \sqrt{6}\,\hbar$.

Assess: The angular momentum must be an integer multiple of $\hbar$ in the Bohr model, but not in the full quantum mechanical model.

P29.25. Prepare: Use $L = \sqrt{l(l+1)}$.

$$\frac{(4.7\times10^{-34}\,\text{J}\cdot\text{s})}{\hbar} = \frac{(4.7\times10^{-34}\,\text{J}\cdot\text{s})}{1.05\times10^{-34}\,\text{J}\cdot\text{s}} = \sqrt{20} = \sqrt{4\cdot5} \;\Rightarrow\; l = 4$$

Solve: With $l = 4$ the smallest n could be is 5.

$$E_5 = -\frac{13.60\,\text{eV}}{5^2} = -0.544\,\text{eV}$$

Assess: This is a typical hydrogen atom energy.

P29.27. Solve: Si, Ge, and Pb are all in the same column of the periodic table as carbon. The electron configuration of Si $(Z = 14)$ is $1s^2 2s^2 2p^6 3s^2 3p^2$. The electron configuration of Ge $(Z = 32)$ is

$$1s^2 2s^2 2p^6 3s^2 3p^6 4s^2 3d^{10} 4p^2$$

The electron configuration of Pb $(Z = 82)$ is

$$1s^2 2s^2 2p^6 3s^2 3p^6 4s^2 3d^{10} 4p^6 5s^2 4d^{10} 5p^6 6s^2 4f^{14} 5d^{10} 6p^2$$

Assess: It is necessary to recall that the $4f$ subshell fills before the $5d$ subshell and after $6s$. In Ge, the $4s$ subshell fills before the $3d$ subshell.

P29.31. Prepare: We are told there is something wrong with each electron configuration; we suspect it is either states in the wrong order or more electrons than permitted in a given state.

Solve:

(a) There can't be eight electrons in a p state.

(b) There can't be three electrons in an s state.

Assess: Our assumption was correct.

P29.35. Prepare: Consult Figure 29.24. We assume the sodium atom starts in the ground state ($3s$). Because of the selection rule $\Delta l = \pm 1$ only p states can be excited directly, so the photon must excite the atom to the $4p$ state first and allow emission to get down to the $4s$.

Solve: The $4p$ state is 3.76 eV above the ground state.

$$\lambda = \frac{1240 \text{ eV} \cdot \text{nm}}{\Delta E_{\text{atom}}} = \frac{1240 \text{ eV} \cdot \text{nm}}{3.76 \text{ eV}} = 330 \text{ nm}$$

Assess: Don't forget the selection rule.

P29.37. Solve: (a) A $4p \rightarrow 4s$ transition is allowed because $\Delta l = 1$. Using the sodium energy levels from Figure 29.24, the wavelength is

$$\lambda = \frac{hc}{\Delta E} = \frac{1240 \text{ eV nm}}{3.76 \text{ eV} - 3.19 \text{ eV}} = 2180 \text{ nm}$$

(b) A $3d \rightarrow 4s$ transition is not allowed because $\Delta l = 2$ violates the selection rule that requires $\Delta l = 1$.

Assess: Remember your selection rules!

P29.41. Prepare: Since $1.00 \text{ mW} = 1.00 \times 10^{-3}$ W, the laser emits $E_{\text{light}} = 1.00 \times 10^{-3}$ J of light energy per second. This energy consists of N photons.

Solve: The energy of each photon is

$$E_{\text{photon}} = hf = \frac{hc}{\lambda} = \frac{(6.63 \times 10^{-34} \text{ J} \cdot \text{s})(3 \times 10^{8} \text{ m/s})}{633 \times 10^{-9} \text{ m}} = 3.14 \times 10^{-19} \text{ J}$$

Because $E_{\text{light}} = NE_{\text{photon}}$, the number of photons is

$$N = \frac{E_{\text{light}}}{E_{\text{photon}}} = \frac{1.00 \times 10^{-3} \text{ J}}{3.14 \times 10^{-19} \text{ J}} = 3.18 \times 10^{15}$$

So, photons are emitted at the rate $3.18 \times 10^{15} \text{ s}^{-1}$.

Assess: As the laser emits considerable energy per second, we expected the emission of a large number of photons.

P29.43. Prepare: If we divide the pulse energy, which is 500 mJ, by the photon energy, which we can calculate using $E_{\text{photon}} = hf = \dfrac{hc}{\lambda}$, we get the number of photons in each pulse of light.

Solve: (a) The energy of each photon is

$$E_{\text{photon}} = hf = \frac{hc}{\lambda} = \frac{(6.63 \times 10^{-34} \text{ J} \cdot \text{s})(3.00 \times 10^{8} \text{ m/s})}{690 \times 10^{-9} \text{ m}} = 2.88 \times 10^{-19} \text{ J}$$

The number of photons in each pulse of light is

$$N = \frac{E_{\text{total}}}{E_{\text{photon}}} = \frac{0.500 \text{ J}}{2.88 \times 10^{-19} \text{ J}} = 1.7 \times 10^{18}$$

(b) The *rate* of emission, that is, the number of photons per second is

$$R = \frac{1.74 \times 10^{18}}{10^{-8} \text{ s}} = 1.7 \times 10^{26} \text{ photons/s}$$

Assess: Each pulse of energy 0.500 J packs in a large number of photons.

P29.47. Prepare: For a neutral atom, the number of electrons is the same as the number of protons, which is the atomic number Z. An atom's mass number is $A = Z + N$, where N is the number of neutrons. Mass of the proton is 1.67×10^{-27} kg and its volume is $4\pi/3(7.0 \times 10^{-15}\,\text{m})^3$.

Solve: (a) For a ^{197}Au atom, $Z = 79$. So, $N = 197 - 79 = 118$. A neutral ^{197}Au atom contains 79 protons, 79 electrons, and 118 neutrons.

(b) Assuming that the neutron rest mass is the same as the proton rest mass, the density of the gold nucleus is

$$\rho_{\text{nucleus}} = \frac{197\, m_{\text{proton}}}{\frac{4\pi}{3}(7.00 \times 10^{-15}\,\text{m})^3} = \frac{197(1.67 \times 10^{-27}\,\text{kg})}{\frac{4\pi}{3}(7.00 \times 10^{-15}\,\text{m})^3} = 2.29 \times 10^{17}\,\text{kg/m}^3$$

(c) The nuclear density in part **(b)** is 2.01×10^{13} times the density of lead.

Assess: The mass of the matter is primarily in the nuclei and the volume of the matter is essentially due to the electrons around the nuclei.

P29.49. Prepare: The mass of an atom is concentrated almost entirely in the nucleus.

Solve: The volumes of the nucleus and the atom are

$$V_{\text{nucleus}} = \frac{4\pi}{3}\left(\frac{1.0 \times 10^{-14}\,\text{m}}{2}\right)^3 = 0.524 \times 10^{-42}\,\text{m}^3 \qquad V_{\text{atom}} = \frac{4\pi}{3}\left(\frac{1.2 \times 10^{-10}\,\text{m}}{2}\right)^3 = 9.05 \times 10^{-31}\,\text{m}^3$$

$$\Rightarrow \frac{V_{\text{nucleus}}}{V_{\text{atom}}} = \frac{0.524 \times 10^{-42}\,\text{m}^3}{9.05 \times 10^{-31}\,\text{m}^3} \approx 5.8 \times 10^{-13}$$

Thus $5.8 \times 10^{-11}\% = 0.000000000058\%$ of the atom's volume contains all the mass. The percent of empty space in an atom is 99.999999999942%.

P29.51. Prepare: The nucleus of an atom is very small and it contains protons and neutrons. $1\,\text{fm} = 10^{-15}$ m.

Solve: (a) The electric force between two protons in the nucleus is

$$F_{\text{E}} = \frac{1}{4\pi\varepsilon_0}\frac{e^2}{(2.0\,\text{fm})^2} = \frac{(8.99 \times 10^9\,\text{N} \cdot \text{m}^2/\text{C}^2)(1.60 \times 10^{-19}\,\text{C})^2}{(2.0 \times 10^{-15}\,\text{m})^2} = 58\,\text{N}$$

(b) The gravitational force between two protons in the nucleus is

$$F_{\text{G}} = \frac{Gm^2}{(2.0\,\text{fm})^2} = \frac{(6.67 \times 10^{-11}\,\text{N} \cdot \text{m}^2/\text{kg}^2)(1.67 \times 10^{-27}\,\text{kg})^2}{(2.0 \times 10^{-15}\,\text{m})^2} = 4.7 \times 10^{-35}\,\text{N}$$

Because $F_{\text{G}} \ll F_{\text{E}}$, gravitational force could not be the force to hold two protons together.

P29.57. Prepare: The four lines in the absorption spectrum imply a total of five levels. We ignore any selections rules.

Solve: The possible emissions are then $5 \to 4, 5 \to 3, 5 \to 2, 5 \to 1, 4 \to 3, 4 \to 2, 4 \to 1, 3 \to 2, 3 \to 1, 2 \to 1$. This is a total of 10.

Assess: The electron could go from level 5 to 3, for instance, and then from 3 to 2 and then from 2 to 1.

P29.59. Prepare: See Figure 29.19.

Solve: Equation 29.17 gives the wavelengths in the hydrogen spectrum:

$$\lambda_{n \to m} = \frac{\lambda_0}{\left(\frac{1}{m^2} - \frac{1}{n^2}\right)} \qquad m = 1, 2, 3, \ldots \; n = m+1, m+2, \ldots$$

For the Lyman series, $m = 1$ and $n = 2, 3, \ldots$ The shortest possible wavelength of the Lyman series corresponds to $m = 1$ and $n \to \infty$. That is, $\lambda_{\infty \to 1} = \lambda_0 = 91.1$ nm.

P29.61. Prepare: Photons are absorbed in a quantum jump from a lower energy level to a higher energy level. The energy of the emitted photon is exactly equal to the energy between the starting and the ending levels. The energy levels of the stationary states of the hydrogen atom are $E_n = -13.60 \text{ eV}/n^2$ and $n = 1, 2, 3,...$ In the ground state $(n = 1)$, $E_1 = -13.60$ eV.

Solve: The change in energy when the hydrogen atom absorbs a 12.75 eV photon is

$$E_n - E_1 = 12.75 \text{ eV} \Rightarrow (-13.60 \text{ eV})\left(\frac{1}{n^2} - 1\right) = 12.75 \text{ eV} \Rightarrow \frac{1}{n^2} = 1 - \frac{12.75}{13.60} = 0.0625 \Rightarrow n = 4$$

When the atom, having been excited to $n = 4$, undergoes a quantum jump to the next lowest energy level (corresponding to $n = 3$), the emitted wavelength is given by Equation 29.2:

$$\lambda_{4 \to 3} = \frac{91.1 \text{ nm}}{1/3^2 - 1/4^2} = 1870 \text{ nm}$$

Assess: This wavelength is in the infrared region.

P29.65. Prepare: We will use the energy-level diagram of the hydrogen atom in Figure 29.18 for energies in the $n = 1$ and $n = 3$ states.

Solve: (a) Nearly all atoms spend nearly all of their time in the ground state $(n = 1)$. To cause an emission from the $n = 3$ state, the electrons must excite hydrogen atoms from the $n = 1$ state to the $n = 3$ state. The energy gained by each atom is

$$E_3 - E_1 = -1.51 \text{ eV} - (-13.60 \text{ eV}) = 12.09 \text{ eV} \approx 12.1 \text{ eV}$$

This means the electrons must each lose 12.09 eV of kinetic energy. Thus the electrons must have *at least* $K_{\min} = 12.09 \text{ eV}$ of kinetic energy to cause the emission of 656 nm light. The minimum speed of the electrons is

$$v_{\min} = \sqrt{\frac{2K_{\min}}{m_{\text{elec}}}} = \sqrt{\frac{2(12.09 \text{ eV})}{9.11 \times 10^{-31} \text{ kg}} \times \frac{1.6 \times 10^{-19} \text{ J}}{1 \text{ eV}}} = 2.06 \times 10^6 \text{ m/s}$$

(b) An electron gains 12.1 eV of kinetic energy by being accelerated through a potential difference of 12.1 V. This is simply the definition of *electron volt*.

P29.67. Prepare: The energy and wavelength of the light emitted are related by $E = hc/\lambda$. The energy-level diagram for the hydrogen atom is shown in Figure 29.19. The selection rule for transitions is $\Delta l = \pm 1$.

Solve: The energy associated with this wavelength of light is determined by

$$E = hc/\lambda = (6.63 \times 10^{34} \text{ J} \cdot \text{s})(3.0 \times 10^8 \text{ m/s})(1\text{eV}/(1.6 \times 10^{-19} \text{ J}))/(6.56 \times 10^{-7} \text{ m}) = 1.89 \text{eV}$$

Checking the energy-level diagram in Figure 29.19, note that this energy is associated with the $n = 3$ to the $n = 2$ transition. Applying the selection rule $\Delta l = \pm 1$ to this transition, we have the following transitions: $3s \to 2p, 3p \to 2s,$ and $3d \to 2p$.

Assess: These three transitions will all give up the same energy and wavelength of light.

P29.71. Prepare: From Figure 29.29 and the accompanying text we see that the energy decrease of 2.10 eV means the transition was from the $3p$ state to the $3s$ state. The $3p$ state has orbital angular momentum quantum number $l = 1$ and the $3s$ state has $l = 0$.

Solve: The equation for the orbital angular momentum, in terms of l is $L = \sqrt{l(l+1)}\hbar$. The change in L is

$$\Delta L = \sqrt{0(0+1)}\hbar - \sqrt{1(1+1)}\hbar = -\sqrt{2}\hbar$$

In other words, it changes by $\sqrt{2}\hbar$ and it is a decrease.

Assess: This line is the bright yellow line in the spectrum of sodium.

P29.73. Prepare: We need to use the condition $\Delta l = 1$ to determine the allowed transitions of the emission spectrum, and the equation

$$E = \frac{hc}{\lambda} = \frac{1240 \text{ eV} \cdot \text{nm}}{\lambda}$$

to find the wavelengths. Visible wavelengths are in the range 400–700 nm, with infrared being greater than 700 nm and ultraviolet being less than 400 nm. The absorption spectrum has *only* transitions from the ground state, so the wavelengths of the emission spectrum that are in the absorption spectrum belong to transitions that end on the $2s$ ground state.

Solve:

Transition	(a) Wavelength	(b) Type	(c) Absorption
$2p \rightarrow 2s$	670 nm	VIS	Yes
$3s \rightarrow 2p$	816 nm	IR	No
$3p \rightarrow 2s$	324 nm	UV	Yes
$3p \rightarrow 3s$	2696 nm	IR	No
$3d \rightarrow 2p$	611 nm	VIS	No
$3d \rightarrow 3p$	24800 nm	IR	No
$4s \rightarrow 2p$	498 nm	VIS	No
$4s \rightarrow 3p$	2430 nm	IR	No

NUCLEAR PHYSICS

Q30.3. Reason: (a) No. With $Z = 3$ and $N = 27$ neutrons this would lie too far away from the line of stability shown in Figure 30.3.
(b) No. For large Z, $N > Z$ for stability. In the given case we would have $N = Z = 92$ and this, too, would lie too far from the line of stability (on the other side, in contrast with part **(a)**).
Assess: The line of stability closely follows the $N = Z$ line for small but tends toward larger N (that is, $N > Z$) as Z increases.

Q30.5. Reason: This is because the isotopes have the same electronic structure and will therefore emit practically the same spectrum. However, they do have different masses, which is a feature that can be detected in a mass spectrometer.
Assess: The emitted light from an atom, measurable with an ordinary spectrometer, tells you about its electrons. The mass, measurable with a mass spectrometer, tells you about the nucleus (since the nucleus has most of the mass of an atom).

Q30.7. Reason: It is clear from the graph that 1/4 of the original nuclei are left after 8 days. In general, 1/4 are left after two half-lives; therefore two half-lives equals 8 days, so one half-life must be 4 days.
Assess: There is no need to do any fancy math on this question. We can see that there would be 500,000 nucleui left when $t = 4\,\text{d}$.

Q30.9. Reason: We must realize that when there are 750 B nuclei there are 250 A nuclei. Clearly there are 1/4 of the original A nuclei left, and that takes two half-lives, so $t = 2 \times 10\,\text{s} = 20\,\text{s}$.

Assess: We can also do this mathematically by starting with Equation 30.12 and solving for t.

$$N = N_0 \left(\frac{1}{2}\right)^{t/t_{1/2}}$$

$$\frac{N}{N_0} = \left(\frac{1}{2}\right)^{t/t_{1/2}}$$

$$\frac{t}{t_{1/2}} = \log_{1/2}\left(\frac{N}{N_0}\right)$$

$$t = (t_{1/2})\log_{1/2}\left(\frac{N}{N_0}\right)$$

Use the change-of-base formula.

$$t = (t_{1/2})\frac{\ln\left(\frac{N}{N_0}\right)}{\ln\left(\frac{1}{2}\right)} = (10\,\text{s})\frac{\ln\left(\frac{250}{1000}\right)}{\ln\left(\frac{1}{2}\right)} = 20\,\text{s}$$

Rather than use the change-of-base formula, you could also realize that you need to raise 1/2 to the second power to get 1/4 so that $\log_{1/2}(250/1000) = 2$.

Q30.15. Reason: The unknown X may be identified as follows:

(a) $^{222}_{86}\text{Rn} \rightarrow {}^{218}_{84}\text{Po} + {}^{4}_{2}X$; X must be an alpha particle, which is a He atom nucleus $^{4}_{2}\text{He}$.

(b) $^{228}_{88}\text{Ra} \rightarrow {}^{228}_{89}\text{Ac} + {}^{0}_{-1}X$; X must be an electron or beta-minus radiation.

(c) $^{140}_{54}\text{Xe} \rightarrow {}^{140}_{55}\text{Cs} + {}^{0}_{-1}X$; X must be an electron or beta-minus radiation.

(d) $^{64}_{29}\text{Cu} \rightarrow {}^{64}_{28}\text{Ni} + {}^{0}_{1}X$; X must be a positron or beta-plus radiation.

Assess: The equation is balanced by conserving mass and charge.

Q30.17. Reason: (a) This nucleus is in an excited state and as it decays the proton in the higher level will drop down to the lower level. The number of protons and neutrons will remain the same and the nucleus will get rid of the excess energy by emitting a gamma ray.
(b) This nucleus is in a stable unexcited state.
(c) This nucleus is in an excited state and as it decays the neutron in the higher level will drop down to the empty lower level proton state. That is, a neutron will become a proton and an electron as follows: $n \rightarrow p + {}^{0}_{-1}e + \text{energy}$. During this process a beta-minus is emitted.
Assess: Nuclear energy–level diagrams provide a lot of information about the nucleus.

Q30.19. Reason: As you radiate the apple, you are putting energy into it. This energy will break molecular bonds and cause the molecules to vibrate. These vibrations will cause the temperature of the apple to rise. After some time, the apple will appear almost as if it had been heated in the oven.
Assess: The process is somewhat like a microwave oven. Microwaves are just like gamma rays except for the amount of energy.

Q30.21. Reason: In a gamma scan, phosphorus-containing molecules with an attached ^{99}Tc atom (a gamma emitter) are administered to the patient. The gamma emitting molecules are absorbed preferentially by the tumor so that the tumor is highlighted. In an x-ray, the tumor may not show up because it may look like the neighboring tissues. In this way, the gamma scan actually shows bodily function and not just structure.
Assess: X-ray images only show structure because they merely show whether an x-ray was able to penetrate the bodily tissue or not. By contrast the gamma scan shows which parts of the body are using certain molecules.

Q30.25. Reason: The patient is given a dose of a phosphorous compound labeled with the gamma emitter ^{99}Tc. We have been informed that ^{99}Tc is taken up and retained by cells in the liver and spleen. If a patient has a tumor that destroys these cells, no ^{99}Tc will be in the region of the tumor. This will show up as a dark spot on the monitor and film.
Assess: Simple ideas frequently contribute a lot to humankind.

Q30.27. Reason: A proton consists of two up quarks and one down quark while a neutron consists of one up quark and two down quarks. When the proton absorbs the electron, one of the up quarks must turns into a down quark so that the proton turns into a neutron. Also, a neutrino is emitted. The reaction can be written as an equation:

$$u + e^{-} \rightarrow d + \nu_e$$

Assess: This seems reasonable in light of the equation in the book for beta-plus decay:

$$u \rightarrow d + e^{+} + \nu_e$$

The only difference between this equation and the one for electron capture is that electron capture has an electron on the left and beta-plus decay has a positron on the right. It is as if the electron has been "subtracted" from both sides in the electron capture equation. In this way of viewing things, the positron is like the "negative" of an electron.

Q30.31. Reason: After one half-life it will be half as bright; after two half-lives it will be one-quarter as bright; after three half-lives it will be one-eighth as bright. So to be only one-tenth as bright it must be slightly older than three half-lives. $3 \times 12.3 \text{ y} = 36.9 \text{ y}$, so the answer is just longer than that, or 40 years. The correct choice is C.

Assess: Fifty years would be too old, as it would be more than four half-lives and the source would be less than one-sixteenth as bright.

Q30.33. Reason: The isotope with the longer half-life takes longer to decay, so there will be fewer decays per unit of time. Hence, the ^{89}Sr sample with the shorter half-life will have the higher activity. The answer is A.

Assess: It makes sense that the one that lasts longer has fewer decays per year.

Problems

P30.1. Prepare: The leading superscript on the elemental notation is the mass number A. The nucleus is composed of Z protons and $A - Z$ neutrons.

Solve: (a) ^{3}H has $Z = 1$ proton and $3 - 1 = 2$ neutrons.

(b) ^{40}Ar has $Z = 18$ protons and $40 - 18 = 22$ neutrons.

(c) ^{40}Ca has $Z = 20$ protons and $40 - 20 = 20$ neutrons.

(d) ^{239}Pu has $Z = 94$ protons and $239 - 94 = 145$ neutrons.

P30.3. Prepare: Using Appendix D, we can take a weighted average of the atomic masses of the different isotopes to find the chemical atomic mass of lithium.

Solve: The isotope ^{6}Li has atomic mass 6.015121 u and an abundance of 7.50% The isotope ^{7}Li has atomic mass 7.016003 u and an abundance of 92.50%. The weighted average is

$$(6.015121 \text{ u})(0.0750) + (7.016003 \text{ u})(0.9250) = 6.94 \text{ u}$$

The chemical atomic mass is 6.94 u.

Assess: A number close to 7, but a little less, is reasonable since most lithium atoms have an atomic mass of around 7 but some have an atomic mass of around 6.

P30.9. Prepare: We can use the equation $B = \Delta m \cdot c^2$ to find the binding energy. The quantity Δm is the extra mass you would have if the protons, neutrons and electrons of the atom were made free particles. The superscript on the element is the mass number, A The nucleus is composed of Z protons and $A - Z$ neutrons.

Solve: For ^{3}He:

$$\Delta m = Zm_\text{H} + Nm_\text{n} - m_\text{atom} = 2(1.007825 \text{ u}) + 1(1.008665 \text{ u}) - 3.016029 \text{ u} = 0.008286 \text{ u}$$

The binding energy is

$$B = \Delta m \cdot c^2 = (0.008286 \text{ u})\left(\frac{931.49 \text{ MeV}/c^2}{1 \text{ u}}\right)c^2 = 7.718 \text{ MeV} \approx 7.72 \text{ MeV}$$

and the binding energy per nucleon is $(1/3)(7.718 \text{ MeV}) = 2.57 \text{ MeV}$.

For ^{4}He:

$$\Delta m = Zm_\text{H} + Nm_\text{n} - m_\text{atom} = 2(1.007825 \text{ u}) + 2(1.008665 \text{ u}) - 4.002602 \text{ u} = 0.030378 \text{ u}$$

The binding energy is

$$B = \Delta m \cdot c^2 = (0.030378 \text{ u})\left(\frac{931.49 \text{ MeV}/c^2}{1 \text{ u}}\right)c^2 = 28.297 \text{ MeV} \approx 28.3 \text{ MeV}$$

and the binding energy per nucleon is $(1/4)(28.297 \text{ MeV}) = 7.07 \text{ MeV}$. Because the binding energy per nucleon is more for ^{4}He than for ^{3}He, ^{4}He is more tightly bound.

Assess: The binding energies per nucleon are consistent with Figure 30.5.

P30.11. Prepare: We can use the equation $B = \Delta m \cdot c^2$ to find the binding energy. The quantity Δm is the extra mass you would have if the protons, neutrons and electrons of the atom were made free particles. The superscript on the element is the mass number, A. The nucleus is composed of Z protons and $A - Z$ neutrons.

Solve: (a) For ^{14}N:

$$\Delta m = Zm_{\mathrm{H}} + Nm_{\mathrm{n}} - m_{\mathrm{atom}} = 7(1.007825 \text{ u}) + 7(1.008665 \text{ u}) - 14.003074 \text{ u} = 0.112356 \text{ u}$$

The binding energy is $B = \Delta m \cdot c^2 = (0.112356 \text{ u})\left(\dfrac{931.49 \text{ MeV}/c^2}{1 \text{ u}}\right)c^2 = 104.7 \text{ MeV}$ and the binding energy per

nucleon is $(1/14)(104.7 \text{ MeV}) = 7.48 \text{ MeV}$.

(b) For ^{56}Fe:

$$\Delta m = Zm_{\mathrm{H}} + Nm_{\mathrm{n}} - m_{\mathrm{atom}} = 26(1.007825 \text{ u}) + 30(1.008665 \text{ u}) - 55.934940 \text{ u} = 0.528460 \text{ u}$$

The binding energy is $B = \Delta m \cdot c^2 = (0.528460 \text{ u})\left(\dfrac{931.49 \text{ MeV}/c^2}{1 \text{ u}}\right)c^2 = 492.3 \text{ MeV}$ and the binding energy per

nucleon is $(1/56)(492.3 \text{ MeV}) = 8.79 \text{ MeV}$.

(c) For ^{207}Pb :

$$\Delta m = Zm_{\mathrm{H}} + Nm_{\mathrm{n}} - m_{\mathrm{atom}} = 82(1.007825 \text{ u}) + 125(1.008665 \text{ u}) - 206.975871 \text{ u} = 1.748904 \text{ u}$$

The binding energy is $B = \Delta m \cdot c^2 = (1.748904 \text{ u})\left(\dfrac{931.49 \text{ MeV}/c^2}{1 \text{ u}}\right)c^2 = 1629 \text{ MeV}$ and the binding energy per

nucleon is $(1/207)(1629 \text{ MeV}) = 7.87 \text{ MeV}$.

Assess: The graph of binding energy per nucleon has a broad maximum at $A \approx 60$, consistent with the given values.

P30.13. Prepare: We can use the equation $B = \Delta m \cdot c^2$ to find the binding energy. The quantity Δm is the extra mass you would have if the protons, neutrons, and electrons of the atom were made free particles. The superscript on the element is the mass number, A The nucleus is composed of Z protons and $A - Z$ neutrons.

Solve: For ^{240}Pu:

$$\Delta m = Zm_{\mathrm{H}} + Nm_{\mathrm{n}} - m_{\mathrm{atom}} = 94(1.007825 \text{ u}) + 146(1.008665 \text{ u}) - 240.053808 \text{ u} = 1.946832 \text{ u}$$

The binding energy is $B = \Delta m \cdot c^2 = (1.946832 \text{ u})\left(\dfrac{931.49 \text{ MeV}/c^2}{1 \text{ u}}\right)c^2 = 1813.45 \text{ MeV}$ and the binding energy per

nucleon is $(1/240)(1813.45 \text{ MeV}) = 7.56 \text{ MeV}$.

For ^{133}Xe:

$$\Delta m = Zm_{\mathrm{H}} + Nm_{\mathrm{n}} - m_{\mathrm{atom}} = 54(1.007825 \text{ u}) + 79(1.008665 \text{ u}) - 132.905906 \text{ u} = 1.201179 \text{ u}$$

The binding energy is $B = \Delta m \cdot c^2 = (1.201179 \text{ u})\left(\dfrac{931.49 \text{ MeV}/c^2}{1 \text{ u}}\right)c^2 = 1118.89 \text{ MeV}$ and the binding energy per

nucleon is $(1/133)(1118.89 \text{ MeV}) = 8.41 \text{ MeV}$.

Assess: These values are consistent with Figure 30.6.

P30.15. Prepare: We can use the equation $B = \Delta m \cdot c^2$ to find the binding energy. The quantity Δm is the extra mass you would have if the protons, neutrons, and electrons of the atom were made free particles. The superscript on the element is the mass number, A. The nucleus is composed of Z protons and $A - Z$ neutrons.

Solve: (a) First the reactants: For ^{3}H:

$$\Delta m = Zm_{\mathrm{H}} + Nm_{\mathrm{n}} - m_{\mathrm{atom}} = 1(1.007825 \text{ u}) + 2(1.008665 \text{ u}) - 3.106049 \text{ u} = 0.009106 \text{ u}$$

The binding energy is $B = \Delta m \cdot c^2 = (0.009106 \text{ u})\left(\dfrac{931.49 \text{ MeV}/c^2}{1 \text{ u}}\right)c^2 = 8.4821 \text{ MeV}.$

For ^{3}He:

$$\Delta m = Zm_{\text{H}} + Nm_{\text{n}} - m_{\text{atom}} = 2(1.007825 \text{ u}) + 1(1.008665 \text{ u}) - 3.016029 \text{ u} = 0.008286 \text{ u}$$

The binding energy is $B = \Delta m \cdot c^2 = (0.008286 \text{ u})\left(\dfrac{931.49 \text{ MeV}/c^2}{1 \text{ u}}\right)c^2 = 7.7183 \text{ MeV}.$

The binding energy per nucleon is the total binding energy divided by the total number of nucleons $(1/6)(8.4821 \text{ MeV} + 7.7183 \text{ MeV}) = 2.700 \text{ MeV}.$

Now the products: For ^{2}H:

$$\Delta m = Zm_{\text{H}} + Nm_{\text{n}} - m_{\text{atom}} = 1(1.007825 \text{ u}) + 1(1.008665 \text{ u}) - 2.014102 \text{ u} = 0.002388 \text{ u}$$

The binding energy is $B = \Delta m \cdot c^2 = (0.002388 \text{ u})\left(\dfrac{931.49 \text{ MeV}/c^2}{1 \text{ u}}\right)c^2 = 2.2244 \text{ MeV}.$

For ^{4}He:

$$\Delta m = Zm_{\text{H}} + Nm_{\text{n}} - m_{\text{atom}} = 2(1.007825 \text{ u}) + 2(1.008665 \text{ u}) - 4.002602 \text{ u} = 0.030378 \text{ u}$$

The binding energy is $B = \Delta m \cdot c^2 = (0.030378 \text{ u})\left(\dfrac{931.49 \text{ MeV}/c^2}{1 \text{ u}}\right)c^2 = 28.2968 \text{ MeV}.$

The binding energy per nucleon is the total binding energy divided by the total number of nucleons $(1/6)(2.2244 \text{ MeV} + 28.2968 \text{ MeV}) = 5.089 \text{ MeV}.$

(b) The difference in the binding energies per nucleon is $5.089 \text{ MeV} - 2.700 \text{ MeV} = 2.389 \text{ MeV}$

Assess: The products have a higher binding energy per nucleon than the reactants; this is expected.

P30.21. Prepare: Equation 30.6 gives the general scheme for alpha decay.

$$^A_Z X \rightarrow \,^{A-4}_{Z-2}Y + \alpha + \text{energy}$$

So the daughter has a smaller atomic number by two.

Solve:

$$^{238}_{94}\text{Pu} \rightarrow \,^{234}_{92}\text{U} + \alpha + \text{energy}$$

So the daughter is ^{234}U.

Assess: ^{234}U is naturally rare; most of it is produced by alpha decay of $^{238}_{94}$Pu. It is also radioactive.

P30.23. Solve: (a) The decay is $X \rightarrow \,^{224}\text{Ra} + \alpha$, so $X = \,^{228}$Th.

(b) The decay is $X \rightarrow \,^{207}\text{Pb} + e^-$, so $X = \,^{207}$Tl.

(c) The decay is $^7\text{Be} + e^- \rightarrow X$, so $X = \,^7$Li.

(d) The decay is $X \rightarrow \,^{60}\text{Ni} + \gamma$, so $X = \,^{60}$Ni.

P30.25. Prepare: The decay is $^{228}\text{Th} \rightarrow \,^{224}\text{Ra} + \,^4\text{He}$. The masses of the nuclei are given in Appendix D, and we note that $m(^{228}\text{Th}) = 228.028715 \text{ u}$, $m(^{224}\text{Ra}) = 224.020186 \text{ u}$, and $m(^4\text{He}) = 4.002602 \text{ u}$.

Solve: The energy released is

$$E = \Delta mc^2 = (228.028716 \text{ u} - 224.020187 \text{ u} - 4.002602 \text{ u})c^2$$

$$= (0.005927 \text{ u})c^2 \times \frac{931.49 \text{ MeV}/c^2}{1 \text{ u}} = 5.52 \text{ MeV}$$

Assess: Essentially all of this energy goes into the alpha particle's kinetic energy.

P30.29. Prepare: The outcome of beta-minus decay is described by Equation 30.9. In beta-minus decay, Z increases by 1 and A remains constant. We can use this information to find the daughter nucleus. We can use Equation 30.12 to find the fraction of tritium which remains after 20 years.

Solve: **(a)** In the beta-minus decay, Z increases from 1 to 2 but A remains 3. Thus the daughter nucleus is ^{3}He. The decay can be written symbolically as: $^3\text{H} \rightarrow {}^3\text{He} + e^- + \text{energy}$.

(b) The fraction of tritium remaining after 20 years can be obtained from Equation 30.12:

$$N = N_0\left(\frac{1}{2}\right)^{t/t_{1/2}} \Rightarrow \frac{N}{N_0} = \left(\frac{1}{2}\right)^{20/12.3} = 0.32$$

Thus 32% of the tritium remains.

Assess: Since 20 years is between one and two half lives of tritium, it is reasonable that the fraction remaining after that time is between 50% and 25%.

P30.33. Prepare: From Equation 30.16, the ratio of the activity of a sample at time t to its activity at time 0 is $(1/2)^{t/t_{1/2}}$. We can set this equal to the 1% and solve for t.

Solve:

$$\left(\frac{1}{2}\right)^{t/t_{1/2}} = 0.01 \Rightarrow (t/t_{1/2})\ln(1/2) = \ln(0.01)$$

$$\Rightarrow t = \frac{\ln(100)}{\ln(2)}t_{1/2} = \frac{\ln(100)}{\ln(2)}(8.0\text{ d}) = 53\text{ d}$$

After 53 d, the activity is reduced to 1% of its original value.

Assess: This looks reasonable since 53 days is about seven half lives and $(1/2)^7 = 1/128$ which is about 1%.

P30.35. Prepare: We can use Equation 30.16 to find the activity of a sample after time t has passed. The only difficulty is that the time and the half-life are in different units so a conversion factor will be necessary.

Solve:

$$R = R_0\left(\frac{1}{2}\right)^{t/t_{1/2}} = (0.10\text{ mCi})\left(\frac{1}{2}\right)^{\left(\frac{1\text{ yr}}{65\text{ d}}\right)\left(\frac{365\text{ d}}{1\text{ yr}}\right)} = (1.0\times10^{-4}\text{ Ci})(0.0204) = 2.0\ \mu\text{Ci}$$

The activity after one year will be $2.0\ \mu\text{Ci}$.

Assess: Since the half-life is a little more than two months, in one year, a little less than six half-lives have passed. After six half-lives, the original activity is divided by a factor of 64. According to our result, the activity has been divided by about 50 which is a little less than 64. So our result is reasonable.

P30.37. Prepare: From Appendix D, we can find radium's half-life and we can calculate the number of nuclei in a sample of 1.0 g. Then by using Equation 30.15, we can find its activity.

Solve: The half-life of ^{226}Ra is 1600 yr. Since this is a two-significant-figure problem, we can just use 226 as the atomic mass. We know that one mole of ^{226}Ra has a mass of 226 g and contains Avogadro's number of atoms so the number of atoms in the sample is

$$N = (1.0\text{ g})\left(\frac{1\text{ mol}}{226\text{ g}}\right)\left(\frac{6.02\times10^{23}}{1\text{ mol}}\right) = 2.664\times10^{21}$$

We have used an extra significant figure for added accuracy. Now we use Equation 30.15 to find the activity of the sample:

$$R = \frac{0.693N}{t_{1/2}} = \frac{0.693(2.664\times10^{21})}{1600\text{ yr}}\left(\frac{1\text{ yr}}{3.16\times10^7\text{ s}}\right) = 3.7\times10^{10}\text{ Bq}$$

This activity is defined as 1 Ci so one gram of ^{226}Ra has an activity of one Curie. Apparently the Curie is named after the activity of one gram of ^{226}Ra.

Assess: This unit makes it easy to estimate the activity of a sample if you know its half-life. One gram of radium is 1/226 mol. So 1/226 mol of a substance with a half-life of 1600 yr has an activity of 1 Ci. Now many radioactive substances such as ^{238}U have an atomic mass close to that of ^{226}Ra so that one gram of them has about the same number of particles as one gram of ^{226}Ra. Thus we can estimate their activity merely by comparing their half-lives to 1600 yr. For example, to one significant figure, ^{238}U has a life of 4×10^9 yr, so one gram of ^{238}U would have an activity of roughly $\left(\dfrac{1600}{4\times10^9}\right)$ Ci $= 4\ \mu$Ci. (The half-life of ^{238}U is in the denominator since activity is inversely proportional to half-life.)

P30.39. Prepare: We can use Equation 30.16 to relate time passed to initial and final activity.
Solve:

$$R = R_0 e^{-t/\tau} \Rightarrow \ln(R/R_0) = -t/\tau \Rightarrow \tau = -t/\ln(R/R_0) \Rightarrow$$

$$t_{1/2} = \tau \ln 2 = -t\ln 2/\ln(R/R_0) = -(48\ \text{h})\ln 2/\ln\left(\frac{120{,}000\ \text{Bq}}{370{,}000\ \text{Bq}}\right) = 30\ \text{h}$$

The half-life of the sample is 30 h.
Assess: The activity is divided by about three after 48 hr After one half life, a sample's half-life is divided by two and after two half-lives by four, so 48 hr should be between one and two half-lives. Put another way, the half-life of the sample should be between half of 48 hr and 48 hr itself. So 30 hr is a reasonable answer.

P30.43. Prepare: Table 30.4 tells us the natural background exposure in a year is 3.0 mSv.
Solve:

$$\# \text{ hours} = \frac{\text{background exposure}}{\text{extra exposure per hour of flight}} = \frac{3.0\ \text{mSv}}{5\ \mu\text{Sv/h}} = 600\ \text{h}$$

Assess: This demonstrates that flying is fairly safe with regard to extra radiation. People who live at high altitudes get more radiation all the time than those living at sea level.

P30.47. Solve: For α – particles, the RBE $= 20$. The dose in rems of the alpha radiation is $30\ \text{rad} \times 20 = 600\ \text{rem}$. For γ-rays, the RBE $= 1$. The does in rads of 600 rem of γ-rays is 600 rad.

P30.51. Prepare: To find the dose and dose equivalent, we need to know the energy received in the first month. From the activity of the strontium, we can find the number of beta particles produced. We can multiply this number by the energy absorbed per beta particle to obtain the total energy absorbed and then divide by the mass of the skeleton to obtain the dose.

Solve: The number of beta particles produced in the first month is

$$N = R\Delta t = (3.7\times10^5\ \text{Bq})(1\ \text{mo})\left(\frac{30\ \text{d}}{1\ \text{mo}}\right)\left(\frac{24\ \text{h}}{1\ \text{d}}\right)\left(\frac{3600\ \text{s}}{1\ \text{hr}}\right) = 9.59\times10^{11}$$

Half the energy of each beta particle, $(2.8\ \text{MeV})/2 = 1.4\ \text{MeV}$ is absorbed, so the total energy absorbed is

$$N(1.4\ \text{MeV}) = (9.59\times10^{11})(1.4\times10^6\ \text{eV})\left(\frac{1.60\times10^{-19}\ \text{J}}{1\ \text{eV}}\right) = 0.215\ \text{J}$$

To get the dose, we divide the energy absorbed by the mass of the skeleton which is $(17\%)(75\ \text{kg}) = 12.8\ \text{kg}$. The dose is $(0.215\ \text{J})/(12.8\ \text{kg}) = 17\ \text{mGy}$. Finally, since beta particles have an RBE of 1, the dose equivalent is 17 mSv.
Assess: The dose equivalent here is comparable to that in Problem 30.50 because, while the exposure time is greater here by $(1\ \text{mo})/(1\ \text{hr}) = 720$, here the whole mass of the person, 75 kg, absorbs the radiation while in Problem 30.50, only the thyroid absorbed the radiation. The mass is greater here by $(75\ \text{kg})/(0.15\ \text{kg}) = 500$ which has the same order of magnitude as 720. (Also important are the energy absorbed per decay and the activity of the radioactive sample, but their combined effect has the same order of magnitude in Problem 30.50.)

P30.53. Prepare: We will assume that the tracer stays in the body for the life of the tracer. The initial activity of the source is $R_0 = 30\ \mu\text{Ci} = 30\times10^{-6}\ \text{Ci} \times 3.7\times10^{10}\ \text{Bq}/(1\ \text{Ci}) = 1.11\times10^6\ \text{Bq}$.
Solve: The decay rate is

$$r = \frac{1}{\tau} = \frac{\ln 2}{t_{1/2}} = \frac{\ln 2}{5.0\ \text{d} \times 24\ \text{h/d} \times 3600\ \text{s/h}} = 1.605\times10^{-6}\ \text{s}^{-1}$$

The number of radioactive atoms is

$$N_0 = \frac{R_0}{r} = \frac{1.11 \times 10^6 \text{ Bq}}{1.605 \times 10^{-6} \text{ s}^{-1}} = 6.92 \times 10^{11}$$

The number that decay in one week is

$$N_0 - N = N_0 \left(1 - \left(\frac{1}{2}\right)^{t/t_{1/2}} \right) = \left(6.92 \times 10^{11} \right) \left(1 - \left(\frac{1}{2}\right)^{7.0 \text{ d}/5.0 \text{ d}} \right) = 4.30 \times 10^{11}$$

Since 10% of the beta particles escape, there are $0.90 N_0$ disintegrations in the body. Since the average energy of the beta particle, is 0.35 MeV, the energy absorbed by the body is

$$E = 0.35 \text{ MeV} \times (1.60 \times 10^{-19} \text{ J/eV}) \times 0.90 \times (4.30 \times 10^{11}) = 0.00217 \text{ J}$$

Thus, the radiation dose is

$$\frac{0.0217 \text{ J}}{75 \text{ kg}} = 2.9 \times 10^{-4} \text{ J/kg} \times \frac{1 \text{ Gy}}{1 \text{ J/kg}} = 0.29 \text{ mGy}$$

The RBE for beta particles is 1, so the dose equivalent is 0.29 mSv.

P30.55. Prepare: Both an electron and a positron have a rest energy of 0.511 MeV and each pion has a rest energy of 140 MeV. The total energy of an object is its rest energy and kinetic energy $(E = E_0 + K)$.

Solve:

$$E_{0_{\text{electron}}} + E_{0_{\text{positron}}} + 2K = E_{0_{\pi^+}} + E_{0_{\pi^-}}$$

Solving for K and inserting values obtain

$$K = (1/2)(E_{0_{\pi^+}} + E_{0_{\pi^-}} - E_{0_{\text{electron}}} - E_{0_{\text{positron}}})$$

Inserting values

$$K = (1/2)(140 \text{ MeV} + 140 \text{ MeV} - 0.511 \text{ MeV} - 0.511 \text{ MeV}) = 139 \text{ MeV}$$

Assess: The total energy of an object is its rest energy and kinetic energy.

P30.59. Prepare: The activity of a radioactive sample is $R = rN$ and $r = 1/\tau = 0.693/t_{1/2}$. From Appendix D, the half-life of a ^{133}Cs sample is $t_{1/2} = 30 \text{ years} \times 3.15 \times 10^7 \text{ s}/(1 \text{ year}) = 9.45 \times 10^8 \text{ s}$.

Solve: The decay rate is

$$r = \frac{1}{\tau} = \frac{0.693}{t_{1/2}} = \frac{0.693}{9.45 \times 10^8 \text{ s}} = 7.33 \times 10^{-10} \text{ s}^{-1}$$

$$\Rightarrow N = \frac{R_0}{r} = \frac{2.0 \times 10^8 \text{ Bq}}{7.33 \times 10^{-10} \text{ s}^{-1}} = 2.73 \times 10^{17} \text{ atoms}$$

That is, at $t = 0$ s, we have 2.7×10^{17} ^{137}Cs atoms and after a very long time all will have decayed. Thus 2.7×10^{17} beta particles will have been emitted.

P30.61. Prepare: The data give a ^{14}C/^{12}C ratio of $100/1.0 \times 10^{15} = 1.0 \times 10^{-13}$. The ratio in living things is 1.3×10^{-12}.

Solve:

$$t = \frac{t_{1/2}}{\ln(1/2)} \ln\left(\frac{N}{N_0}\right) = \frac{5730 \text{ y}}{\ln(1/2)} \ln\left(\frac{1.0 \times 10^{-13}}{1.3 \times 10^{-12}}\right) = 21{,}000 \text{ y}$$

Assess: This is a reasonable number, both archeologically and scientifically, and well within the range of validity of radiocarbon dating.

P30.63. Prepare: The number of ^{40}K atoms decays exponentially with time. Suppose the number of ^{40}K atoms is N_0 at the time the lava solidifies. There are no ^{40}Ar atoms at that time. As time passes, some ^{40}K decay into ^{40}Ar, but the total number of atoms locked inside the lava is unchanged.

Solve: (a) Call N_K the remaining number of ^{40}K atoms, so the fraction that decays is $1 - N_K/N_0$.

$$1 - \frac{N_K}{N_0} = 1 - \left(\frac{1}{2}\right)^{t/t_{1/2}} = 1 - \left(\frac{1}{2}\right)^{1.8\times10^6 \text{ yr}/1.28\times10^9 \text{ yr}} = 0.00097 \approx 0.10\%$$

(b) $N_{Ar} + N_K = N_0$, where N_K is the remaining number of ^{40}K atoms. Dividing by N_K, we have

$$\frac{N_{Ar}}{N_K} + 1 = \frac{N_0}{N_K} \Rightarrow \frac{N_{Ar}}{N_K} = \frac{N_0}{N_K} - 1 = \left(\frac{1}{2}\right)^{-1.8\times10^6 \text{ yr}/1.28\times10^9 \text{ yr}} - 1 = 0.00098$$

Assess: 1.8 million years isn't much compared to the long half-life, so we don't expect much argon yet. Since the argon comes from decayed potassium we expect the ratios from parts (a) and (b) to be almost identical.

P30.69. Prepare: Because the $\text{RBE} = 1$, the dose is 30 mrad or

$$30 \text{ mrad} = 30\times10^{-3} \text{ rad} \times \frac{0.01 \text{ J/kg}}{1 \text{ rad}} = 30\times10^{-5} \text{ J/kg}$$

Solve: Since only 1/4 of the body received x-ray exposure, the total amount of energy received is

$$30\times10^{-5} \text{ J/kg} \times \frac{1}{4} \times 60 \text{ kg} = 4.5\times10^{-3} \text{ J}$$

The energy of each photon is

$$10 \text{ keV} \times \frac{1.6\times10^{-19} \text{ J}}{1 \text{ eV}} = 1.6\times10^{-15} \text{ J}$$

Thus, the number of x-ray photons absorbed by the body is

$$\frac{4.5\times10^{-3} \text{ J}}{1.6\times10^{-15} \text{ J}} = 2.8\times10^{12}$$

MODERN PHYSICS

PptVII.9. Reason: The two fragments fly off at high speeds, and their kinetic energy is what is eventually converted to electric energy by a turbine and generator. The answer is D.
Assess: The released neutrons also carry away some energy, but not as much as the two large fragments.

PptVII.11. Reason: Because of conservation of momentum (see previous problem) we know the Br nucleus is going faster than the La nucleus (because the Ba is lighter). In the formula for kinetic energy the speed is squared, so the nucleus with the higher speed has the greatest kinetic energy when the momenta have the same magnitude. The answer is A.
Assess: We saw this idea back in the chapter on conservation of momentum.

PptVII.15. Reason: The radiation detected by Detector 5 was not bent by the magnetic field; this means it isn't charged. Gamma radiation is electromagnetic radiation (photons) and not charged. The correct choice is D.
Assess: Charged particles are deviated by a magnetic field, but gamma rays (and all other electromagnetic radiation is not.

PptVII.17. Reason:
(a) The equation for relativistic kinetic energy is $K = (\gamma - 1)E_0$. Solve for v. The mass of an electron is $0.511\,\text{MeV}/c^2$.

$$\frac{K}{mc^2} + 1 = \gamma = \frac{1}{\sqrt{1 - \frac{v^2}{c^2}}}$$

$$v = c\sqrt{1 - \left(\frac{1}{\frac{K}{mc^2}+1}\right)^2} = c\sqrt{1 - \left(\frac{1}{\frac{19\,\text{keV}}{511\,\text{keV}}+1}\right)^2} = 0.265c = 8.0\times10^7 \text{ m/s}$$

(b)

$$(15\text{ MBq})\left(\frac{1\text{ decay/s}}{1\text{ Bq}}\right)(19\text{ keV/decay})\left(\frac{1.6\times10^{-19}\text{ J}}{1\text{ eV}}\right) = 4.56\times10^{-8}\text{ W} \approx 4.6\times10^{-8}\text{ W}$$

(c) The activity decays exponentially.

$$R = R_0\left(\frac{1}{2}\right)^{t/t_{1/2}} = 15\text{ MBq}\left(\frac{1}{2}\right)^{5\,y/12\,y} = 11\text{ MBq}$$

Assess: The 11 MBq in the last part seems reasonable after less than half of a half-life.

PptVII.19. Reason: First express the speed of the ship in m/s.

$$(0.12)(3.0\times10^8 \text{ m/s}) = 3.6\times10^7 \text{ m/s}.$$

(a) For a non-relativistic calculation we use

$$K = \frac{1}{2}mv^2 = \frac{1}{2}(1.7\times10^6 \text{ kg})(3.6\times10^7 \text{ m/s})^2 = 1.1\times10^{21} \text{ J}$$

(b) The number of reactions to produce twice this kinetic energy (since only half of the fusion energy ends up as kinetic energy of the ship) is

$$N = \frac{(2)(1.1\times10^{21} \text{ J})}{18 \text{ MeV}}\left(\frac{1 \text{ eV}}{1.6\times10^{-19} \text{ J}}\right) = 7.65\times10^{32} \text{ reactions} \approx 7.7\times10^{32} \text{ reactions}$$

(c) We need 7.65×10^{32} of each kind of nucleus. The mass of a ^2_1H nucleus is $(2)(1.67\times10^{-27} \text{ kg}) = 3.34\times10^{-27}$ kg and the mass of a ^3_2He nucleus is $(3)(1.67\times10^{-27} \text{ kg}) = 5.01\times10^{-27}$ kg.

$$(7.65\times10^{32})(3.34\times10^{-27} \text{ kg}) = 2.6\times10^6 \text{ kg of } {}^2_1\text{H}$$
$$(7.65\times10^{32})(5.01\times10^{-27} \text{ kg}) = 3.8\times10^6 \text{ kg of } {}^3_2\text{H}$$

Assess: We would need to haul around (or collect from space) a few million kilograms of hydrogen and helium.

PptVII.21. Reason: If we follow the discussion in the text and assume the muons are traveling at a speed close to the speed of light then the time needed to reach the ground in the earth's reference frame is $\Delta t = (120 \text{ km})/(3.0\times10^8 \text{ m/s}) = 400\,\mu s$.
Next we find γ for the 10 GeV muons.

$$\gamma = \frac{K}{E_0} + 1 = \frac{10 \text{ GeV}}{105 \text{ MeV}} + 1 = 96.24$$

In the muons' reference frame the time interval (proper time) is

$$\Delta\tau = \frac{\Delta t}{\gamma} = \frac{400\,\mu s}{96.24} = 4.16\,\mu s$$

This is $(4.16\,\mu s)/(1.5\,\mu s) = 2.77$ half-lives, so the number reaching the ground is

$$N = N_0\left(\frac{1}{2}\right)^{2.77} = 15000$$

Assess: Many more muons reach the ground than expected without time dilation. From the other reference frame we could say the atmosphere is length contracted and arrive at the same answer.